AF358625

Advanced Structural Health Monitoring: From Theory to Applications

Advanced Structural Health Monitoring: From Theory to Applications

Editors

Hugo Rodrigues
Ivan Duvnjak

MDPI • Basel • Beijing • Wuhan • Barcelona • Belgrade • Manchester • Tokyo • Cluj • Tianjin

Editors
Hugo Rodrigues
University of Aveiro
Portugal

Ivan Duvnjak
University of Zagreb
Croatia

Editorial Office
MDPI
St. Alban-Anlage 66
4052 Basel, Switzerland

This is a reprint of articles from the Special Issue published online in the open access journal *Applied Sciences* (ISSN 2076-3417) (available at: https://www.mdpi.com/journal/applsci/special_issues/Structural_Health_Monitoring_Theory_Applications).

For citation purposes, cite each article independently as indicated on the article page online and as indicated below:

LastName, A.A.; LastName, B.B.; LastName, C.C. Article Title. *Journal Name* **Year**, *Volume Number*, Page Range.

ISBN 978-3-0365-5035-0 (Hbk)
ISBN 978-3-0365-5036-7 (PDF)

© 2022 by the authors. Articles in this book are Open Access and distributed under the Creative Commons Attribution (CC BY) license, which allows users to download, copy and build upon published articles, as long as the author and publisher are properly credited, which ensures maximum dissemination and a wider impact of our publications.
The book as a whole is distributed by MDPI under the terms and conditions of the Creative Commons license CC BY-NC-ND.

Contents

About the Editors

Hugo Rodrigues

Hugo Rodrigues is an Associate Professor at the Department of Civil Engineering of the University of Aveiro, teaching several topics related to Structural Analysis, Building Pathology, and Rehabilitation. He received his PhD in Civil Engineering from the University of Aveiro. He has experience in seismic analysis, having participated as a team member in research and development projects, as well as specialized consultancy studies ordered by several public institutions and companies regarding the assessment of seismic risk. His primary research interests are Building Rehabilitation, Structural Health Monitoring, and Seismic Safety, including experimental and numerical activities.

Ivan Duvnjak

Ivan Duvnjak is an Associate Professor at the Faculty of Civil Engineering of the University of Zagreb and Head of Structural Testing Laboratory at the same faculty. He teaches several topics dealing with theory of elasticity and plasticity and mechanical properties of materials. He received his PhD in Civil Engineering from the University of Zagreb. He is co-author of the scholarly book "Theory elasticity and plasticity with methods of problem solving". His primary research interests cover damage detection of structures, structural health monitoring, nondestructive testing, static and dynamic testing of structures, proof loading of bridges, mechanical properties of materials, model updating of structures, etc. He is involved in many national and international research projects, and he is providing consultancy ordered by companies.

Preface to "Advanced Structural Health Monitoring: From Theory to Applications"

Structural Health Monitoring (SHM) is a strategic tool for the monitoring and noninvasive assessment of the health state of existing structures' infrastructures and systems and can provide decision support for reducing operational costs and risks throughout their life cycle. In that sense, this Special Issue presents six research papers and one review paper that deal with the recent developments in the theoretical, computational, experimental, and practical aspects of this field. The Special Issue covers the following topics: sensors for structural health monitoring; algorithms for damage detection and characterization; structural warning systems; model-based methods for predicting structural service life; the application of SHM for various exceptional loads; the influence of environmental and operational conditions; innovative sensing solutions for SHM; cultural heritage damage detection and health monitoring; bridge damage detection and health monitoring; case study applications; and short-term monitoring systems for diagnostic load testing of structures.

Hugo Rodrigues and Ivan Duvnjak
Editors

Editorial

Editorial on the Special Issue: Advanced Structural Health Monitoring: From Theory to Applications

Hugo Rodrigues [1,*] **and Ivan Duvnjak** [2]

1   RISCO, Civil Engineering Department, University of Aveiro, 3810-193 Aveiro, Portugal
2   Department for Engineering Mechanics, Faculty of Civil Engineering, University of Zagreb, 10000 Zagreb, Croatia; ivan.duvnjak@grad.unizg.hr
*   Correspondence: hrodrigues@ua.pt

Citation: Rodrigues, H.; Duvnjak, I. Editorial on the Special Issue: Advanced Structural Health Monitoring: From Theory to Applications. *Appl. Sci.* **2021**, *11*, 11401. https://doi.org/10.3390/app112311401

Received: 23 November 2021
Accepted: 29 November 2021
Published: 2 December 2021

Publisher's Note: MDPI stays neutral with regard to jurisdictional claims in published maps and institutional affiliations.

Copyright: © 2021 by the authors. Licensee MDPI, Basel, Switzerland. This article is an open access article distributed under the terms and conditions of the Creative Commons Attribution (CC BY) license (https://creativecommons.org/licenses/by/4.0/).

This editorial focuses on the interesting studies published within the present Special Issue related to Advanced Structural Health Monitoring. Moreover, it highlights different techniques and approaches for the structural health assessment of Civil Engineering Structures and reports some interesting studies and their outcomes in this specific field.

Structural Health Monitoring (SHM) is a strategic tool for the monitoring and non-invasive assessment of the health state of existing structures' infrastructures and systems and can be applied in several areas, such as aeronautics, mechanical, civil, and electrical engineering. During their lives, systems are exposed to several actions and environmental conditions that can lead to structural and nonstructural damage. Recent advances in sensor technology and techniques have allowed us to gain insight into the diagnosing of material degradation and structural and nonstructural damages.

Nowadays, there is a trend of increasing the service life of structures. Structures are commonly assessed periodically based on the results of visual inspection or local limited nondestructive testing methods. Although visual inspections are essential, the results can often lead to subjective conclusions; therefore, structural health monitoring is essential as a tool that can detect degradation continuously at an early stage of their occurrence. SHM can provide decision support for reducing operational costs and risks throughout their life cycle.

In this Special Issue, six research papers and one review paper deal with the recent developments in the theoretical, computational, experimental, and practical aspects of this field and aim to cover the following topics: sensors for structural health monitoring; algorithms for damage detection and characterization; structural warning systems; model-based methods for predicting the structural service life; the application of SHM for various exceptional loads; the influence of environmental and operational conditions; innovative sensing solutions for SHM; cultural heritage damage detection and health monitoring; bridge damage detection and health monitoring; case study applications; and short-term monitoring systems for diagnostic load testing of structures.

The review paper is focused on the assessment methods and damage detection technologies for existing masonry structures, discussing the traditional methods and the new technologies with a special focus on unmanned aerial vehicles, as well as photogrammetry and close-range remote sensing as a technology that can complement traditional ways of assessment. In addition, the authors presented an example of a graphical interpretation of a case study after an earthquake. This tool may be very useful, especially in large and dispersed affected areas, where it is difficult to have a first assessment of the damage distribution [1].

Vibration-based damage detection in structures to interpret the structures' changes in dynamic properties is a common technique, and Duvnjak et al. [2] proposed a new Mode Shape Damage Index (MSDI) based on the difference between modified modal displacements in the undamaged and damaged state of the structure. The MSDI method can be used to detect the presence of damage, identify single and/or multiple damage locations, and distinguish damage of different severity. It must be emphasized that the severity of damage detected by this method depends on the boundary conditions and location of the damage. The method was tested based on experimental modal analysis on a reinforced concrete plate under different damage cases.

Wang et al. [3] presented a scale physical model test to study the plane gate vibration and holding force under the conditions of the fixed gate opening and closing processes, considering the gate vibration, holding force, and the failure of gate-closing in the closing process and the correlation between the parameters, considering the vibration properties and the correlation of the changes in the different system properties.

Mobaraki et al. [4] presented and validated the application of the observability technique for the structural system identification of 2D models. The main highlight of this work is the changes in variables proposed to linearize the system of equations, in order to shift the non-linear problem and impossible to solve to a linearized system of equations.

Two papers are related to the SHM proposals that can be applied in large infrastructures. The first one proposes an intelligent judgment method for improving the sensitivity of analyzing mechanical parameters of prestressed cables based on the digital twin based on the information collected on large-span prestressed cables by field sensors [5]. The other paper proposes a method for monitoring the structural health of concrete bridges using a condition index by the analytical hierarchy process based on eight indices that are scored based on the experts' views. The method was implemented in several case studies to prove application of the method as an easy-to-use optimization tool in health monitoring and prioritizing programs [6].

Finally, another paper studies the effect of particle size in the pH measurement of cement-based materials, a topic related to a characteristic that is gaining importance in the analysis of cement-based materials, as well as in structural health monitoring and forensic engineering applications. It focuses on the material and structural durability in the context of the structural and non-structural monitorization to ensure that reinforced concrete is highly alkaline to safeguard the passive protective of film for reinforcement of steel bars against corrosion [7].

The topics addressed in the papers are from several fields, including the proposal and validation of new methods, development of techniques to assess the SHM of infrastructures, and methods to evaluate the evolution of chemical material properties.

The editors are confident that the papers reflect significant contributions to the research and development in the various topics addressed. We hope that readers will find all articles of the Special Issue useful and exciting and that the articles will stimulate further research activities in the area of structural health monitoring, sensing and measurement techniques, damage-detection algorithms and characterization, data analysis and structural assessment, and the new and complex solutions and their impacts on the SHM of civil engineering structures.

Acknowledgments: The editors would like to express a special acknowledgement to the authors for their appreciated contributions and to the reviewers for their valuable suggestions that increased the manuscripts' quality.

Conflicts of Interest: The authors declare no conflict of interest.

References

1. Stepinac, M.; Gašparović, M. A Review of Emerging Technologies for an Assessment of Safety and Seismic Vulnerability and Damage Detection of Existing Masonry Structures. *Appl. Sci.* **2020**, *10*, 5060. [CrossRef]
2. Duvnjak, I.; Damjanović, D.; Bartolac, M.; Skender, A. Mode Shape-Based Damage Detection Method (MSDI): Experimental Validation. *Appl. Sci.* **2021**, *11*, 4589. [CrossRef]
3. Wang, Y.; Xu, G.; Li, W.; Liu, F.; Duan, Y. Characteristics of Plane Gate Vibration and Holding Force in Closing Process by Experiments. *Appl. Sci.* **2020**, *10*, 6111. [CrossRef]
4. Mobaraki, B.; Ma, H.; Lozano Galant, J.; Turmo, J. Structural Health Monitoring of 2D Plane Structures. *Appl. Sci.* **2021**, *11*, 2000. [CrossRef]

5. Liu, Z.; Shi, G.; Jiang, A.; Li, W. Intelligent Discrimination Method Based on Digital Twins for Analyzing Sensitivity of Mechanical Parameters of Prestressed Cables. *Appl. Sci.* **2021**, *11*, 1485. [CrossRef]
6. Darban, S.; Ghasemzadeh Tehrani, H.; Karballaeezadeh, N.; Mosavi, A. Application of Analytical Hierarchy Process for Structural Health Monitoring and Prioritizing Concrete Bridges in Iran. *Appl. Sci.* **2021**, *11*, 8060. [CrossRef]
7. Loh, P.; Shafigh, P.; Katman, H.; Ibrahim, Z.; Yousuf, S. pH Measurement of Cement-Based Materials: The Effect of Particle Size. *Appl. Sci.* **2021**, *11*, 8000. [CrossRef]

Review

A Review of Emerging Technologies for an Assessment of Safety and Seismic Vulnerability and Damage Detection of Existing Masonry Structures

Mislav Stepinac [1],* and Mateo Gašparović [2]

[1] Faculty of Civil Engineering, University of Zagreb, 10000 Zagreb, Croatia
[2] Chair of Photogrammetry and Remote Sensing, Faculty of Geodesy, University of Zagreb, 10000 Zagreb, Croatia; mgasparovic@geof.unizg.hr
* Correspondence: mislav.stepinac@grad.unizg.hr

Received: 3 July 2020; Accepted: 21 July 2020; Published: 23 July 2020

Abstract: The construction sector has proven to be one of the slowest sectors to embrace technology—a problem that must be addressed. This problem can be quickly and efficiently addressed in certain aspects of seismic engineering: from seismic risk assessment to damage detection, as well as condition assessments existing structures before or after an earthquake. In this paper, the literature review of assessment methods and damage detection technologies for existing (mainly) masonry structures is presented. Traditional methods are briefly explained, and modern are critically discussed. Special focus is given to unmanned aerial vehicles, as well as, photogrammetry and close-range remote sensing as a technology that can complement traditional ways of assessment and give us data about a structure that is often different to obtain. Graphical interpretation of one post-earthquake case study is provided. Open challenges and opportunities of emerging technologies for faster and easier assessment of seismic safety and vulnerability are presented.

Keywords: earthquake; vulnerability; unmanned aerial vehicle; damage detection; assessment; photogrammetry; close-range remote sensing

---

1. Introduction

Earthquake is an unpredictable and unexpected natural disaster that results in the devastation of both natural and human-made environments. Studying the complexity of an earthquake, including its prediction and its consequences, requires knowledge from various areas of scientific study, ranging from seismology, geology, geodesy, mathematics, and applied statistics all the way to psychology, humanities and social science and of course, structural engineering.

The seismic risk of a structure is a measure of the expected future damage caused by the earthquake, which is expected to occur in the site of construction. It depends on three factors: hazard, vulnerability, and exposure [1,2]. Most casualties from earthquakes are associated with collapsing buildings. In this regard, the continuous assessment and monitoring of the seismic safety and vulnerability of buildings is a challenging task, especially when large-area evaluations are required [3]. The construction sector plays a significant role in the alleviation of the effects and consequences caused by earthquakes and new technologies, products, systems, design approaches are regularly introduced to a broader community [4–12].

It is generally assumed that an earthquake cannot be predicted (although it can be expected) or simulated (although we can approximate it), and that its destructive effects cannot be prevented (although they can be minimized or, at worst, optimized). However, building our infrastructure with the basic state-of-the-art (STAR) principles in mind will provide us with the ability to keep it safe from a predicted level of seismic activity (a detail frequently overlooked) and, more importantly in

---

Appl. Sci. **2020**, *10*, 5060; doi:10.3390/app10155060   www.mdpi.com/journal/applsci

the seismic design to save lives. The importance of seismic design is to protect property and life in buildings. But to have a proper seismic design, we need to learn from existing structures.

The construction sector is the slowest one to appropriate new technologies [13,14]—and this is something that should be changed. This motivates the quest for finding the technological solution for the safety assessment of existing structures. As the least innovative sector, civil engineering must appropriate technological advances from other disciplines in order to be able to keep up with them. This applies to all aspects of the sector, from construction and planning to seismic and safety assessment.

In this paper, the literature review on modern technologies for the assessment of seismic safety and vulnerability of existing structures will be explained. Special focus will be set on unmanned aerial vehicles (UAV) and supporting equipment, as well as, photogrammetry and close-range remote sensing.

2. The Assessment Methods for Condition Inspection and Identification of Seismic Vulnerability of Existing Masonry Buildings

Regarding economy and sustainability, it is of great significance for a society to maintain existing building stock instead demolish and rebuild them. The contribution of the assessment phase to the decision-making process is crucial for the correct assessment, verification, and maintenance of existing buildings [15–17].

The assessment of an existing structure can be done in various stages with increased accuracy, the level of accuracy, thereby depending on the quantity and the quality of available data as well as on the significance of the structure. The developments in the areas of inspection, nondestructive techniques (NDT), structural health monitoring (SHM), and structural analysis of existing structures, together with recent guidelines for reuse and conservation, allow for safer, economical and more adequate remedial measures [18].

The assessment techniques focus mainly on damage identification, damage localization, and damage evaluation, as well as the determination of certain material properties of existing structures. There are several new NDT and semi-NDT methods on the market, different in sophistication and purpose. As the focus of this research is mainly on masonry structures or combined masonry and concrete/timber structures, the most important NDT methods for existing masonry structures can be summarized: visual inspection [19], rebound hammer [20,21], measurement of reinforcement location [22], stress wave transmission [23], ultrasonic velocity testing [24], sonic velocity testing [25,26], surface penetrating radar [25,27], infrared thermography [28], flatjack tests [29,30], damage identification using impact vibrations [31], acoustic emission [32], etc. The basic advantages and disadvantages of the mentioned methods are described in Table 1. More detailed information can be found in [33].

The issue of seismic vulnerability assessment and restoration of underperforming existing buildings is a very significant and difficult problem [34]. To develop efficient assessment methods, it is necessary to draw knowledge and information from various other scientific fields. The seismic vulnerability requires specialized technical skills [35].

The preliminary work on seismic assessment was based only on visual inspection of the buildings [19,36], NDT [15,24,32,37,38], and the experience of an engineer. The last decade has seen a growth in the technological advancement of various instruments that can considerably help in the estimation of structural safety and seismic behavior of existing structures. The applicability of the thermal camera for building diagnostics and detection of energy-related building defects has been investigated by numerous authors [39–44]. The importance close-range remote sensing and using multispectral cameras for damage assessment and crack detection on buildings was emphasized by numerous researchers [45–47]. In the last few years, a growing number of studies have shown the great potential of hyperspectral cameras for detection and mapping cracks on buildings [48,49]. Unmanned aerial vehicles (UAV) can be used for i.e., disaster management [50,51], crack identification using image processing [47,52], seismic vulnerability [53], architectural assessment of heritage sites and structures [54], and structural assessment using image processing techniques [55].

UAV-based photogrammetry [56–58] and LiDAR (Light Detection and Ranging) [59–61] devices are indeed used after earthquakes to scan affected areas, but this method is mostly used for larger areas only. The advantages and disadvantages of the new technologies which should complement traditional methods are given in Table 2. Mentioned techniques are especially proven to be good at a post-earthquake evaluation of the complete urban/rural areas and buildings individually.

Table 1. Basic advantages and disadvantages of NDT and semi-NDT inspection methods for existing masonry structures.

Method	Advantages	Disadvantages
Visual inspection	• Cheap • Fast (or no) preparation • Immediate results • Non-destructive	• Skilled working force needed • Detection of only larger defects • Possible misinterpretation of cracks
Stress wave transmission	• The presence of reinforcements or moisture do not affect the results • Non-destructive	• The results should be used for qualitative purposes • Highly affected by moisture and salt content
Ultrasonic & sonic velocity testing, acoustic emission	• Fast and higher accuracy in results Non-destructive method • Accessibility to just one side of the element needed • Not affected by moisture	• High skilled and educated working force needed • Misreading of signals • Wrong interpretation of data • Transducers must be coupled on the material surface • Small defects can affect results • The wave speed is automatically calculated
Impact echo	• Non-destructive method • Accessibility to just one side of the element needed	• Slow method – requires a lot of measurements to map larger regions • Stress wave energy is reflected at air boundaries
Surface penetrating radar	• Cracks, bonds, delamination does not affect the results • Fast method • Non-destructive method • Possibility to investigate different conditions such as salt content and moisture content, location of reinforcements, deterioration, etc.	• Relatively expensive • Mortar interfaces can mask energy reflected from points of interest
Rebound hammer test	• Ease of use • Relatively cheap • Fast method	• Results are for a local point • The results are not directly related to the strength of surface • Affected by surface and moisture condition • Not completely suitable for masonry (better in assessment of concrete structures) • Not completely NDT – leaves a little hole in the masonry unit
Flatjack system	• Gives compressive strength with reasonable accuracy • Ease of use	• Semi-destructive method • Time-consuming • Drilling and cutting is not always easy (especially for stone masonry) • Requires repair of a mortar joint • Frequent calibration required

Table 2. Basic advantages and disadvantages of new technologies for the rapid assessment of the vulnerability of structures.

Technology/Method	Advantages	Disadvantages
UAV	• Affordable • Fast (or no) preparation • Ease of use • Immediate results • Nondestructive • Constant improvement of technology • Safety—remote control of the vehicle • Available inspection of literally any place of the building	• Limited battery capacity • Depends on weather conditions • Legal restrictions related to drone flight • Limited amount of damage which can be detected
LiDAR	• Data can be collected quickly and with great precision • Can be used during the day and night • Mostly independent of weather conditions • Noncomplex postprocessing • 3D data acquiring	• Unprofitable for detecting small building cracks • Ineffective in heavy rain • Large database that is difficult to interpret • Price • Non or low-quality textures
High-resolution cameras	• Quickly available data • High quality and precision photography	• Price • A large amount of data and sometimes takes a long time to get useful information
Infrared thermography	• Rapid evaluation of big areas • NDT method • Real-time measurements	• High skilled and educated working force needed • Affected by moisture content, texture, or reflections which can lead to inaccurate measurements • Relatively expensive
Photogrammetry	• Nondestructive method • Cheap, simple, and fast data collection process • Very precise • It is possible to measure displacements and deformations of structures without contact • High-resolution photorealistic 3D building models • High-resolution textures • Better for smaller building cracks detection compared to LiDAR	• Not possible without the presence of light (during the day) • Depends on the weather conditions • Vegetation can create problems when taking photos • Large data sets • Complex postprocessing • Required postprocessing to get 3D data
360 cameras	• Fast and efficient technology • Can be mounted on vehicles that are then passed through the area affected by the earthquake. • Possible to create an online platform (such as Google Street View) where earthquake damage can be identified and documented	• If the distances at which the photos were taken are large or the vehicle is moving too fast, there may be a problem connecting the images • Vegetation and certain objects can create problems when taking photos • Inability to detect roof damage • Unprofitable for detecting small building cracks

3. UAV Platform for Multi-Sensor Photogrammetry Aerial Mapping

Nowadays, technology and sensor minimization enable the development of the multi-sensor custom-made UAVs (unmanned aerial vehicles) for multi-sensor aerial mapping. As already mentioned, UAVs can be applied to acquiring spatial data about the condition of buildings [59], cultural heritage [56], but also for acquiring spatial data of larger areas for risk assessment of damage detection and mapping, e.g., after earthquakes [57,58]. These platforms are equipped with the newest sensors for in-flight position determination. These sensors include GNSS (Global Navigation Satellite System), RTK (real-time kinematic), and IMU (inertial measurement unit) systems that allow positioning in centimeter accuracy in real time. Today, significant importance in the development of the UAV platform was devoted to developing new technologies for stabilization and orientation of the multi-sensor system [62–64]. The UAV platforms for multi-sensor aerial mapping can be made based on the open-source or commercial flight controllers and technologies and are able to carry out mapping missions autonomously [65–67]. For high-accuracy photogrammetric measurement of the safety and seismic vulnerability of existing structures, the systems are equipped with sensors that measure the obstacle distance and therefore enable flights at a close distance to buildings (e.g., microradar and/or LiDAR). Furthermore, these systems are the foundation for fast and accurate collection of large amounts of spatial data for building damage detection and mapping. In Figure 1, one scenario of the post-earthquake assessment with a UAV is shown. An engineer cannot see what happened to a building at higher floors or roof of a structure when doing on-site assessment from the street. Very often, the main problems (especially for moderate intensity earthquakes) are chimneys, damaged roof structures, damaged gable walls, etc. With a UAV device, all of the mentioned information can be easily assessed.

(**a**)　　　　　　　　　　　　　　　　(**b**)

Figure 1. *Cont.*

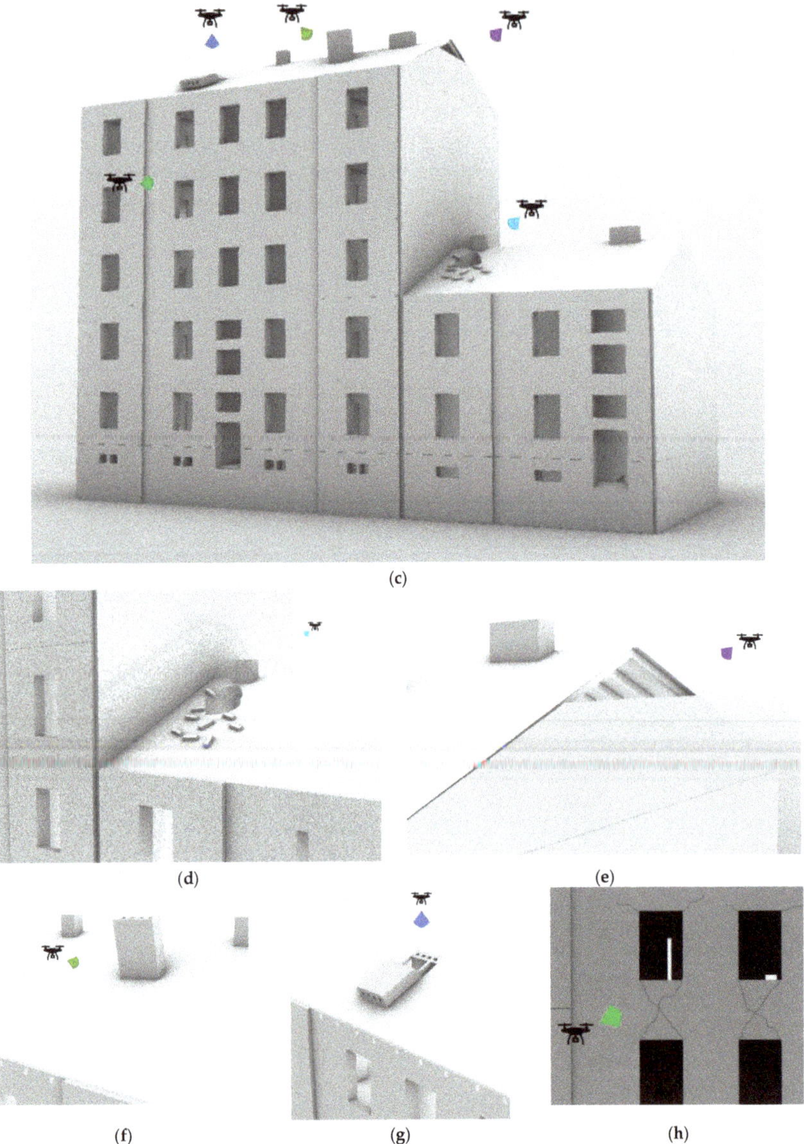

Figure 1. Photogrammetric multi-sensor systems on UAVs enable a fast and accurate collection of large amounts of spatial data for building damage detection and mapping. (**a**) Undamaged building, (**b**) visual inspection from street level, (**c**) post-earthquake assessment with UAV systems, (**d–h**) possible damages on a building after an earthquake.

3.1. Multi-Sensors System for Photogrammetry Data Acquisition

The aforementioned advances in technology and sensor minimization give us the opportunity to the development of unique, highly specialized multi-sensors system for photogrammetry data acquisition. Today, these systems can contain four independent types of cameras: RGB camera, thermal, multispectral, and hyperspectral camera. All these cameras complement each other in spectral, radiometric, and geometric characteristics. The new multi-sensor system is capable for the detection of damage and anomalies on a building that is not visible to the human eye. The importance of using various kinds of different cameras, from thermal to multispectral cameras for damage assessment and crack detection on buildings, was emphasized by numerous researchers [48,52]. Although all these cameras were already used for building diagnostics assessment, so far, a multi-sensor system has not been developed for this purpose. It should be noted that the multi-sensor system should be modular, and it should be able to use different camera lenses, depending on building size and a priori accuracy. The newly built multi-sensor system should enable accurate data acquisition for building damage assessment, and the system harness the potential of all implemented cameras.

After the design and development of the multi-sensor system, it is necessary to develop new methods for calibration, synchronization, and fusion of the collected data. For each camera, the development of the new calibration method based on photogrammetric methods must be obtained. The geometric calibration for internal camera parameters calculation must be done [68,69] and followed by the radiometric calibration for each camera [70–73]. Further, for the autonomous spatial data acquisition from all cameras in a multi-sensor system, the system synchronization is crucial [70,74,75]. Adequately synchronized multi-sensor system are able to autonomously and/or remotely capture images from all cameras at the same time. Once the data has been properly collected, it is necessary to research and develop new imagery fusion methods. The new fusion methods developed for the multi-sensor system with various independent cameras allow the collection of a fused image that contains the data of all the above sensors (visible part of the spectrum, thermal, multispectral and hyperspectral data). The fusion process increases the spatial resolution of final multi-band imagery. The usage of fused, high-resolution, multi-band imagery enables better and more accurate damage detection on buildings.

3.2. Remote Sensing Methods for Automatic Detection and Mapping of Weak Structural Parts and Components

After the multi-sensor system has collected the data, to speed up the process, the mapping of weak structural parts and components should be performed. Such methods allow for the rapid analysis of large amounts of spatial data, which cannot be processed manually. The current development of new methods is based primarily on the analysis of all available data of the multi-sensor system. Taking advantage of individual bands outside the visible spectrum, such as images captured by a thermal, multispectral, and hyperspectral camera, enable detection, highlighting, and mapping of features that are not visible to the human eye [45,46]. The use of fused imagery further enhances the spatial resolution of the imagery and enables the detection of even the smallest cracks in the object [76,77]. The development and use of these methods enable rapid detection and accurate mapping of the smallest cracks that may not be visible on the objects themselves. The application of developed methods significantly accelerates the process of gathering information on the structure and condition of existing buildings.

3.2.1. Photogrammetric Three-Dimensional (3D) Building Modeling Methods

Optimal photogrammetric methods need to be developed for an accurate and measurable high-resolution photorealistic 3D building model based on a multi-sensor system and GeoSLAM system [78]. Photogrammetric methods are primarily based on an adequate phototriangulation procedure for all images. During the process of phototriangulation, the self-calibration of cameras control the stability of their internal elements [71]. The development of methods for the rapid creation of 3D models include the test of sensor accuracy for determining the elements of the external orientation of real-time images (GNSS + RTK, IMU) [62,69]. Markers can be used on the objects to control the phototriangulation process and the final 3D model. Nowadays, indoor spatial data for 3D models are usually acquired with the GeoSLAM system. The final 3D model can be made by the combination of data acquired with the multi-sensor system and GeoSLAM system data. Furthermore, spectralon and other markers to perform radiometric calibration can be placed on the objects [71]. This ensures the accuracy of the spectral signature of the recorded objects. For controlling thermal images, the temperature of the object can be measured by a measuring device and sensors at the time of data collection. This enables obtaining an accurate 3D model in the thermal spectrum, i.e., the accuracy of the temperature obtained by the multi-sensor system. The development and implementation of methods for photogrammetric creation of SfM (Structure-from-motion) point clouds enable the creation of high-resolution, dense 3D point clouds [79–81]. The dense point clouds are filtered by newly developed outlier detection algorithms for gross error points detection (e.g., glass area) [82–84]. For quality and accurate point cloud collection on objects with uniform textures, methods for radiometric enhancement of textural features were developed. Radiometric equalization of photogrammetric images based on new methods allow for the creation of seamless photorealistic 3D models [85,86]. Further, with new 3D modeling methods, the point cloud can be transformed into a high-resolution photorealistic 3D building model (Figure 2).

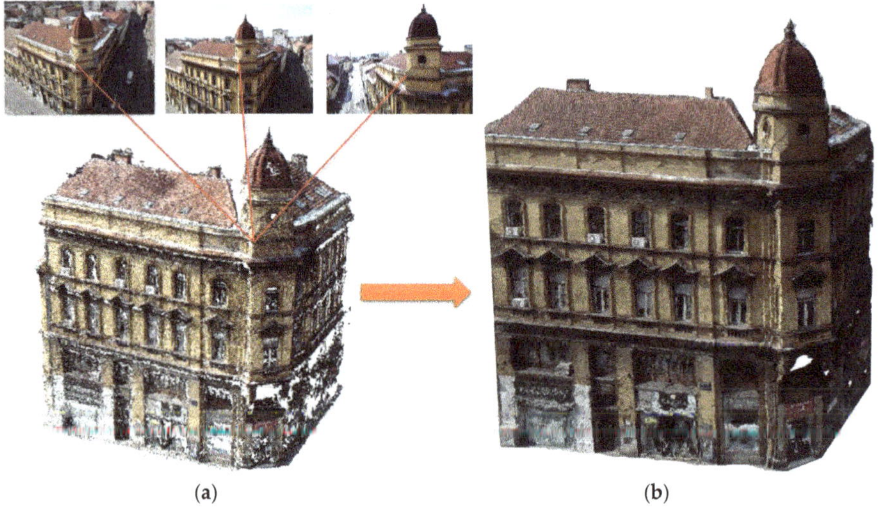

(a)   (b)

Figure 2. Photogrammetric 3D building modeling methods. From (**a**) photogrammetric creation of SfM (Structure-from-motion) point clouds to the (**b**) high-resolution photorealistic 3D building model.

3.2.2. Automatic Methods for Building Damage Detection and Mapping

The photorealistic 3D models and products (e.g., from orthophoto) can be used for the detection and mapping of building damage. The developed methods are based on a combination of unsupervised and supervised image classification methods. Furthermore, the methods can be technologically developed on object-based machine learning algorithms [87–90]. Such an approach allows rapid damage detection on buildings, which is important because of the large amount of high-resolution spatial data. Preliminary research related to the topic of developing automatic methods for the classification has been made [91]. Newly developed methods for the detection and mapping of building damage can use all bands of the multi-sensor system, but also spectral indices. Spectral indices allow further highlighting of certain phenomena on earth or objects [91–93]. The use of spectral indices has been shown to increase the accuracy of classifications. In some cases, spectral indices allow features to be highlighted on an object that cannot be detected on any single band in the visible or out of the visible part of the spectrum. In addition to the spectral indices, the newly developed methods also use the Haralick texture features (Gray Level Co-occurrence Matrix) to detect and emphasize the edge between damaged and undamaged parts on objects [94–98]. Using an object-based approach enables rapid and accurate mapping of damage. After the mapping, the system immediately assigns statistical indicators such as area, depth, etc. (Figure 3). The development of a new automatic method for structural damage detection and mapping is a major scientific contribution in the field of photogrammetry and remote sensing applications in civil engineering.

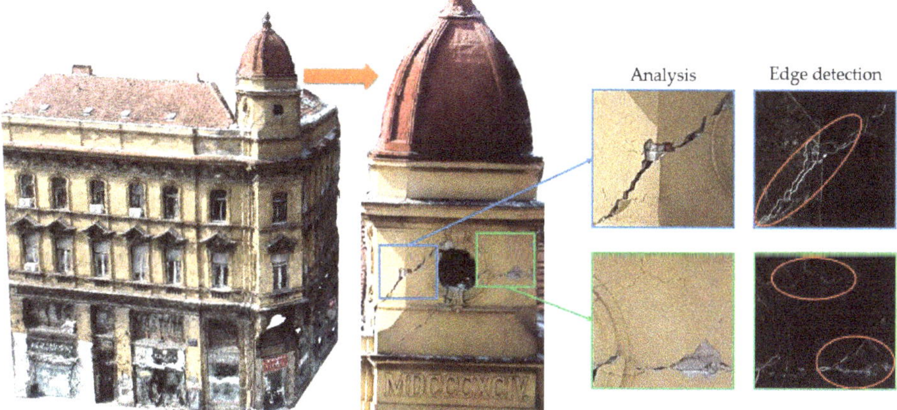

Figure 3. Automatic building damage detection and mapping based on the photogrammetric 3D methods and remote sensing techniques.

4. Discussion and Conclusions

The knowledge of seismic behavior of structures has seen significant advancements in recent decades. Still, there is much left to be considered. We are still unfamiliar with a number of things that must be taken into account, such as the magnitude, direction, and scale of earthquakes themselves, as well as their stochastic (random) nature. Some fundamental decisions still need to be made, e.g., whether to opt for deterministic or probabilistic modeling, static or dynamic load modeling, linear or nonlinear seismic behavior of structures.

Are we able to minimize the effects of an earthquake, or is optimizing them still the only option? How much do we know about the seismic risk or the potential consequences of seismic activity on our infrastructure? How accurate are we in assessing the seismic vulnerability of existing buildings? Alas, most of these questions remain unanswered as more research needs to be done on seismic risk and seismic vulnerability assessment. Concentrating only on visible issues cannot resolve this questions.

Conventional approaches only allow us to assess the condition of an existing structure once its stability has already been compromised. They also only help us when we know from experience that a certain building is vulnerable to seismic activity. But what about any invisible parameters? Are we ready to appropriate new technologies in the engineering sector? Can the use of thermography and UAVs help us expand our abilities in seismic vulnerability assessment? Can any new methods point us to other parameters crucial to the development of this scientific field? How do we precisely characterize data gathered with the help of new technologies? Answering these questions may give way to new principles in the field and foster the use of new technologies of photogrammetry and close-range remote sensing much needed in such a traditional sector as civil engineering.

Insight in material and structural properties obtained by assessments and structural analyses and evaluation can help us is in modeling, analyzing, and predicting the performance of existing structures and is always a base for the preparation of the data for more precise analyses. However, it is not always possible to have a comprehensive on-site assessment for rapid evaluation of buildings.

The basic idea of this paper is to show the possibilities for the development of an easy and fast procedure for seismically generated damage detection of buildings with new technologies such as UAVs and optical inspections. The open questions and possible future trends in the assessment of the seismic vulnerability of the existing structures is also discussed. The seismic vulnerability of a structure is a quantity associated with its weakness in the case of earthquakes of a given intensity so that the value of this quantity and the knowledge of seismic hazard allows us to evaluate the expected damage from future earthquakes [99]. In modern assessment methods, the seismic vulnerability is represented by vulnerability curves, considering, in individual cases only, some structural characteristics of the affected buildings [100]. Throughout history, a variety of methods have been used to assess the vulnerability that can be divided into empirical and analytical (both approaches can also be used in different hybrid methods) and an approach based on engineering judgment by experts. Methods of vulnerability assessments mainly model damage to a discrete scale of damage commonly using three to six categories [101], but still, there are no unified approaches on a European level. High-definition precision images of buildings using satellites, LiDARs, drones, Google Street View and the like can give us information on floor dimensions, height, floors, etc. It is important to emphasize that such information alone is not sufficient, and it needs to be processed by experts, but also supplemented by other data relevant for damage assessments (e.g., structural system, function, and usage).

A compilation of all the data can be obtained to get a unique view of the seismic vulnerability of existing structures. A combination of the data from "traditional" assessments with the results from modern techniques can help us in the identification of weak structural parts and components (faults, big cracks, and openings, short cantilevers, and walls, short columns, soft storeys). With the knowledge of the architectural background and styles of the city, a quick estimation of the period when the building was built can be obtained, i.e., in Croatia, tall windows can tell us that the building was built during the Austro-Hungarian times. This data can relate to construction methods, wooden ceilings, lack of glazing and confiners, greater wall thickness, different material quality, etc. All these important factors are influencing the seismic behavior of the existing structure. With measurement data of foundation settlings, soil stiffness can be assessed or may indicate some differential settling. In addition, the building stiffness can be estimated, which can lead to an estimation of the number of internal walls. Assessment of roofs with UAVs can identify the deformation of roof plates and can serve as a basis for the identification of internal walls, which can also lead to estimation of stiffness. With thermography, different building materials can be identified (masonry vs. concrete) and existing confining elements with their dimensions and location in the structures can be pointed out. With a combination of methods, and estimation of wall thicknesses (with basic geometry) quick interpretation of structural safety can be estimated. Deformation of outer walls obtained by photogrammetry combined with energy losses by thermography and architectural rules of the construction period, and with expert opinions can lead us to a rough estimation of the number of walls (considering the knowledge of load-bearing plate system). All of the mentioned activities can give us the level of seismic vulnerability and safety of the existing structures. The obtained data, new procedures, and techniques can help us in the estimation of seismic risks of bigger regions or building blocks (Figure 4).

A conceptual design of building blocks can be easily identified (weak parts, floors on different heights, and influence of one building to another), which gives the information of seismic risk.

The development of a modern procedure for structural damage detection and estimation of seismic risk and vulnerability must be improved with new and accessible technologies.

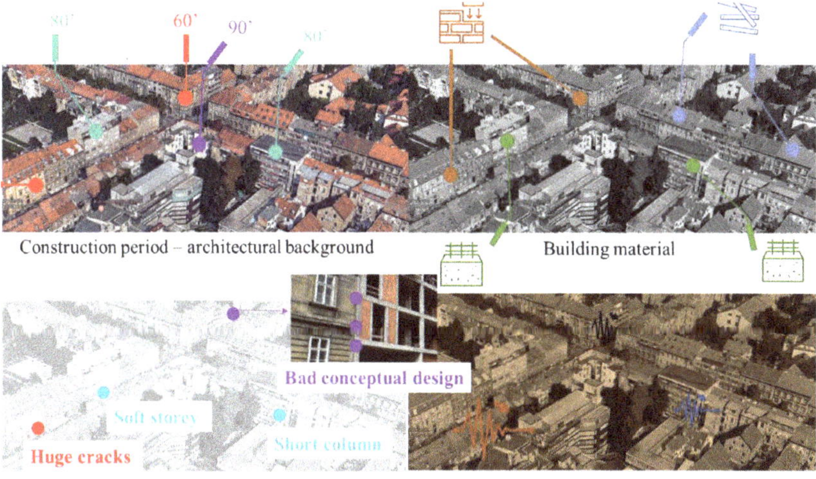

Figure 4. Predicting the seismic vulnerability of buildings and seismic risk of building blocks with the combination of various data.

Author Contributions: Conceptualization, M.S. and M.G.; methodology, M.S. and M.G.; validation, M.S. and M.G.; formal analysis, M.S. and M.G.; investigation, M.S. and M.G.; resources, M.S. and M.G.; photo credit, M.S. and M.G.; writing—original draft preparation, M.S. and M.G.; writing—review and editing, M.S. and M.G.; visualization, M.S. and M.G.; project administration, M.S. and M.G.; funding acquisition, M.S and M.G. All authors have read and agreed to the published version of the manuscript.

Funding: This research was partially funded by Croatian Science Foundation, grant number UIP-2019-04-3749 (ARES project—Assessment and rehabilitation of existing structures—development of contemporary methods for masonry and timber structures). This work was supported by the University of Zagreb for the project: "Advanced photogrammetry and remote sensing methods for environmental change monitoring" (Grant No. RS4ENVIRO).

Conflicts of Interest: The authors declare no conflicts of interest. The funders had no role in the design of the study; in the collection, analyses, or interpretation of data; in the writing of the manuscript, or in the decision to publish the results.

References

1. Barbieri, G.; Biolzi, L.; Bocciarelli, M.; Fregonese, L.; Frigeri, A. Assessing the seismic vulnerability of a historical building. *Eng. Struct.* **2013**, *57*, 523–535. [CrossRef]
2. Atalić, J.; Šavor Novak, M.; Uroš, M. Seismic risk for Croatia: Overview of research activities and present assessments with guidelines for the future. *Appl. Sci.* **2019**, *71*, 923–947.
3. Geiss, C.; Klotz, M.; Taubenb, H. Remote sensing for seismic building vulnerability assessment. In Proceedings of the Second European Conference on Earthquake Engineering and Seismology, Istanbul, Turkey, 25–29 August 2014.
4. Antolinc, D.; Žarni, R.; Stepinać, M.; Rajčič, V.; Krstevska, L.; Tashkov, L. Simulation of earthquake load imposed on timber-glass composite shear wall panel. In *COST Action TU0905 Mid-Term Conference Structural Glass*; Belis, J., Louter, C., Mocibob, D., Eds.; CRC Press: London, UK, 2013.
5. Fragiacomo, M.; Dujic, B.; Sustersic, I. Elastic and ductile design of multi-storey crosslam massive wooden buildings under seismic actions. *Eng. Struct.* **2011**, *33*, 3043–3053. [CrossRef]
6. Pejovic, J.; Jankovic, S. Seismic fragility assessment for reinforced concrete high-rise buildings in Southern Euro-Mediterranean zone. *Bull. Earthq. Eng.* **2016**, *14*, 185–212. [CrossRef]
7. Hancilar, U.; Taucer, F.; Corbane, C. Empirical fragility functions based on remote sensing and field data after the 12 January 2010 Haiti earthquake. *Earthq. Spectra* **2013**, *27*, 157–177. [CrossRef]

8.   Hancilar, U.; Çaktı, E.; Erdik, M. Earthquake performance assessment and rehabilitation of two historical unreinforced masonry buildings. *Bull. Earthq. Eng.* **2012**, *10*, 307–330. [CrossRef]

9.   Mendes, N.; Lourenço, P.B. Sensitivity analysis of the seismic performance of existing masonry buildings. *Eng. Struct.* **2014**, *11*, 143–160. [CrossRef]

10.   Stepinac, M.; Šušteršič, I.; Gavrić, I.; Rajčić, V. Seismic design of timber buildings: Highlighted challenges and future trends. *Appl. Sci.* **2020**, *10*, 1380. [CrossRef]

11.   Bedon, C.; Amadio, C. Enhancement of the seismic performance of multi-storey buildings by means of dissipative glazing curtain walls. *Eng. Struct.* **2017**, *152*, 320–334. [CrossRef]

12.   Kišiček, T.; Renić, T.; Lazarević, D.; Hafner, I. Compressive shear strength of reinforced concrete walls at high ductility levels. *Sustainability* **2020**, *12*, 4434. [CrossRef]

13.   Manyika, J.; Chui, M.; Miremadi, M.; Bughin, J.; George, K.; Wilmott, P.; Dewhurst, M. A Future That Works: Automation, Employment, and Productivity. Available online: https://www.semanticscholar.org/paper/A-future-that-works%3A-automation%2C-employment%2C-and-Manyika-Chui/e3ac9558f18234cdf92b730a7386ff16d446a1af (accessed on 1 January 2017).

14.   McKinsey Global Institute. Global Media Report 2014. Available online: https://www.mckinsey.com/industries/technology-media-and-telecommunications/our-insights/global-media-report-2014 (accessed on 9 September 2014).

15.   Machado, J.S.; Riggio, M.; D'Ayala, D. Assessment of structural timber members by non-and semi-destructive methods. *Constr. Build. Mater.* **2015**, *101*, 1155–1156. [CrossRef]

16.   Stepinac, M.; Rajčić, V.; Barbalić, J. Inspection and condition assessment of existing timber structures. *Gradjevinar* **2017**, *69*, 1380. [CrossRef]

17.   Stepinac, M.; Rajčić, V.; Honfi, D. Condition assessment of timber structures–Quantifying the value of information. In *Tomorrow 's Megastructures, Proceedings of the 40th International Association for Bridge and Structural Engineering Symposium (IABSE), Nantes, Italy, 19–21 September 2018*; Curan Associates Inc.: Red Hook, NY, USA, 2018.

18.   Lourenço, P.B. The ICOMOS methodology for conservation of cultural heritage buildings: Concepts, research and application to case studies. In Proceedings of the International Conference on Preservation, Maintenance and Rehabilitation of Historical Buildings and Structures, Tomar, Portugal, 19–21 June 2014. [CrossRef]

19.   Borri, A.; Corradi, M.; De Maria, A.; Sisti, R. Calibration of a visual method for the analysis of the mechanical properties of historic masonry. *Procedia Struct. Integr.* **2018**, *11*, 418–427. [CrossRef]

20.   Breysse, D.; Martínez-Fernández, J.L. Assessing concrete strength with rebound hammer: Review of key issues and ideas for more reliable conclusions. *Mater. Struct. Constr.* **2014**, *47*, 1589–1604. [CrossRef]

21.   Sýkora, M.; Diamantidis, D.; Holický, M.; Marková, J.; Rózsás, Á. Assessment of compressive strength of historic masonry using non-destructive and destructive techniques. *Constr. Build. Mater.* **2018**, *193*, 196–210. [CrossRef]

22.   Agreed, K.; Klysz, G.; Balayssac, J.P. Location of reinforcement and moisture assessment in reinforced concrete with a double receiver GPR antenna. *Constr. Build. Mater.* **2018**, *188*, 1119–1127. [CrossRef]

23.   Sajid, S.H.; Ali, S.M.; Carino, N.J.; Saeed, S.; Sajid, H.U.; Chouinard, L. Strength estimation of concrete masonry units using stress-wave methods. *Constr. Build. Mater.* **2018**, *163*, 518–528. [CrossRef]

24.   Mesquita, E.; Martini, R.; Alves, A.; Antunes, P.; Varum, H. Non-destructive characterization of ancient clay brick walls by indirect ultrasonic measurements. *J. Build. Eng.* **2018**, *19*, 172–180. [CrossRef]

25.   Martini, R.; Carvalho, J.; Barraca, N.; Arêde, A.; Varum, H. Advances on the use of non-destructive techniques for mechanical characterization of stone masonry: GPR and sonic tests. *Procedia Struct. Integr.* **2017**, *5*, 1108–1115. [CrossRef]

26.   Valluzzi, M.R.; Cescatti, E.; Cardani, G.; Cantini, L.; Zanzi, L.; Colla, C.; Filippo, C. Calibration of sonic pulse velocity tests for detection of variable conditions in masonry walls. *Constr. Build. Mater.* **2018**, *192*, 272–286. [CrossRef]

27.   Wai-Lok Lai, W.; Dérobert, X.; Annan, P. A review of Ground Penetrating Radar application in civil engineering: A 30-year journey from Locating and Testing to Imaging and Diagnosis. *NDT E Int.* **2018**, *96*, 58–78. [CrossRef]

28.   Meola, C. Infrared thermography of masonry structures. *Infrared Phys. Technol.* **2007**, *49*, 228–233. [CrossRef]

29. Parivallal, S.; Kesavan, K.; Ravisankar, K.; Sundram, B.A.; Ahmed, A.K.F. Evaluation of in-situ stress in masonry structures by flat jack technique. In Proceedings of the National Seminar Exhibition on Non-Destructive Evaluation (NDE 2011), Chennai, India, 8–10 December 2011; pp. 8–13.

30. Simões, A.; Gago, A.; Bento, R.; Lopes, M. Flat-Jack Tests on Old Masonry Buildings. In Proceedings of the 15th International Conference on Experimental Mechanics, Porto, Portugal, 22–27 July 2012; Volume 1, p. 3056.

31. Ramos, L.F.; De Roeck, G.; Lourenço, P.B.; Campos-Costa, A. Damage identification on arched masonry structures using ambient and random impact vibrations. *Eng. Struct.* **2010**, *32*, 146–162. [CrossRef]

32. Invernizzi, S.; Lacidogna, G.; Lozano-Ramírez, N.E.; Carpinteri, A. Structural monitoring and assessment of an ancient masonry tower. *Eng. Fract. Mech.* **2018**, *210*, 429–443. [CrossRef]

33. Stepinac, M.; Kisicek, T.; Renić, T.; Hafner, I.; Bedon, C. Methods for the assessment of critical properties in existing masonry structures under seismic loads-the ARES project. *Appl. Sci.* **2020**, *10*, 1576. [CrossRef]

34. Negro, P.; Mola, E. A performance based approach for the seismic assessment and rehabilitation of existing RC buildings. *Bull. Earthq. Eng.* **2017**, *15*, 3349–3364. [CrossRef]

35. Lourenco, P.; Karanikoloudis, G. Seismic behavior and assessment of masonry heritage structures. Needs in engineering judgement and education. *RILEM Tech. Lett.* **2019**, *3*, 114–120. [CrossRef]

36. Piazza, M.; Riggio, M. Visual strength-grading and NDT of timber in traditional structures. *J. Build. Apprais.* **2008**, *3*, 267–296. [CrossRef]

37. Dietsch, P.; Kreuzinger, H. Guideline on the assessment of timber structures: Summary. *Eng. Struct.* **2011**, *33*, 2983–2986. [CrossRef]

38. Croce, P.; Beconcini, M.L.; Formichi, P.; Cioni, P.; Landi, F.; Mochi, C.; Giuri, R. Shear modulus of masonry walls: A critical review. *Procedia Struct. Integr.* **2018**, *11*, 339–346. [CrossRef]

39. Fox, M.; Coley, D.; Goodhew, S.; De Wilde, P. Thermography methodologies for detecting energy related building defects. *Renew Sustain. Energy Rev.* **2014**, *40*, 296–310. [CrossRef]

40. Usamentiaga, R.; Venegas, P.; Guerediaga, J.; Vega, L.; Molleda, J.; Bulnes, F.G. Infrared thermography for temperature measurement and non-destructive testing. *Sensors* **2014**, *14*, 12305–12348. [CrossRef] [PubMed]

41. Kylili, A.; Fokaides, P.A.; Christou, P.; Kalogirou, S.A. Infrared thermography (IRT) applications for building diagnostics: A review. *Appl. Energy* **2014**, *134*, 531–549. [CrossRef]

42. Milovanović, B.; Pečur, I.B. Review of active IR thermography for detection and characterization of defects in reinforced concrete. *J. Imaging* **2016**, *2*, 11. [CrossRef]

43. Carpentier, O.; Chartier, O.; Antczak, E.; Descamp, T.; van Parys, L. Active and quantitative infrared thermography using frequential analysis applied to the monitoring of historic timber structures. In Proceedings of the International Conference on Structural Health Assessment of Timber Structures, Wroclaw, Poland, 11 September 2015; pp. 61–70.

44. Khan, F.; Bolhassani, M.; Kontsos, A.; Hamid, A.; Bartoli, I. Modeling and experimental implementation of infrared thermography on concrete masonry structures. *Infrared Phys. Technol.* **2015**, *69*, 228–237. [CrossRef]

45. Mader, D.; Blaskow, R.; Westfeld, P.; Weller, C. Potential of UAV-Based laser scanner and multispectral camera data in building inspection. *Int. Arch. Photogramm. Remote Sens. Spat. Inf. Sci.-ISPRS Arch.* **2016**, *B1*, 1135–1142. [CrossRef]

46. Valença, J.; Gonçalves, L.M.S.; Júlio, E. Damage assessment on concrete surfaces using multi-spectral image analysis. *Constr. Build. Mater.* **2013**, *40*, 971–981. [CrossRef]

47. Ellenberg, A.; Kontsos, A.; Bartoli, I.; Pradhan, A. Masonry crack detection application of an unmanned aerial vehicle. In Proceedings of the International Conference Computing in Civil and Building Engineering, Orlando, FL, USA, 23–25 June 2014. [CrossRef]

48. Santos, B.O.; Valença, J.; Júlio, E. Detection of cracks on concrete surfaces by hyperspectral image processing. *Autom. Vis. Insp. Mach. Vis. II* **2017**, *188*, 72–79. [CrossRef]

49. Santos, B.O.; Valença, J.; Júlio, E. Automatic mapping of cracking patterns on concrete surfaces with biological stains using hyper-spectral images processing. *Struct. Control Health Monit.* **2019**, *26*. [CrossRef]

50. Khan, A.; Gupta, S.; Gupta, S.K. Multi-hazard disaster studies: Monitoring, detection, recovery, and management, based on emerging technologies and optimal techniques. *Int. J. Disaster Risk Reduct.* **2020**, *47*, 101642. [CrossRef]

51. De Silva, L.O.; de Bandeira, R.A.M.; Campos, V.B.G. Proposal to planning facility location using UAV and geographic information systems in a post-disaster scenario. *Int. J. Disaster Risk Reduct.* **2019**, *36*, 101080. [CrossRef]

52. Kim, H.; Lee, J.; Ahn, E.; Cho, S.; Shin, M.; Sim, S.H. Concrete crack identification using a UAV incorporating hybrid image processing. *Sensors* **2017**, *26*, 2052. [CrossRef] [PubMed]
53. Grazzini, A.; Chiabrando, F.; Foti, S.; Sammartano, G.; Spanò, A. A multidisciplinary study on the seismic vulnerability of St. Agostino Church in Amatrice following the 2016 seismic sequence. *Int. J. Archit. Herit.* **2019**, *3058*, 885–902. [CrossRef]
54. Federman, A.; Santana Quintero, M.; Kretz, S.; Gregg, J.; Lengies, M.; Ouimet, C.; Laliberte, J. UAV photgrammetric workflows: A best practice guideline. In Proceedings of the International Archives of the Photogrammetry, Remote Sensing and Spatial Information Services (ISPRS Archives), 26th International CIPA Symposium, Ottawa, CA, Canada, 28 August–1 September 2017. [CrossRef]
55. Barrile, V.; Candela, G.; Fotia, A. Point cloud segmentation using image processing techniques for structural analysis. In Proceedings of the Remote Sensing and Spatial Information Services (ISPRS Annual Photogrammetry), 2nd International Conference of Geomatics and Restoration, Milan, Italy, 8–10 May 2019. [CrossRef]
56. Achille, C.; Adami, A.; Chiarini, S.; Cremonesi, S.; Fassi, F.; Fregonese, L.; Taffurelli, L. UAV-based photogrammetry and integrated technologies for architectural applications—Methodological strategies for the after-quake survey of vertical structures in Mantua (Italy). *Sensors* **2015**, *15*, 15520–15539. [CrossRef]
57. Gomez, C.; Purdie, H. UAV-based photogrammetry and geocomputing for hazards and disaster risk monitoring–A review. *Geoenviron. Disasters* **2016**, *3*, 23. [CrossRef]
58. Lin, Y.S.; Chuang, R.Y.; Yen, J.Y.; Chen, Y.C.; Kuo, Y.T.; Wu, B.L.; Huang, S.-Y.; Yang, C.-J. Mapping surface breakages of the 2018 Hualien earthquake by using UAS photogrammetry. *Terr. Atmos. Ocean Sci.* **2019**, *30*, 351–366. [CrossRef]
59. Hirose, M.; Xiao, Y.; Zuo, Z.; Kamat, V.R.; Zekkos, D.; Lynch, J. Implementation of UAV localization methods for a mobile post-earthquake monitoring system. In Proceedings of the IEEE Work. Environmental Energy and Structural Monitoring Systems (EESMS 2015), Trento, Italy, 9–10 July 2015. [CrossRef]
60. Hudnut, K.W.; Brooks, B.A.; Scharer, K.; Hernandez, J.L.; Dawson, T.E.; Oskin, M.E.; Arrowsmith, J.R.; Goulet, C.A.; Blake, K.; Boggie, M.A.; et al. Airborne lidar and electro-optical imagery along surface ruptures of the 2019 ridgecrest earthquake sequence, Southern California. *Seismol Res. Lett.* **2020**, *11*, 3082. [CrossRef]
61. Jaydev, P.D.; Gregory, D.; Oussama, K.; Vijay, K. (Eds.) *Experimental Robotics*; Springer: Heidelberg, Germany, 2013; Volume 88, pp. 733–743. [CrossRef]
62. Benassi, F.; Dall'Asta, E.; Diotri, F.; Forlani, G.; Morra di Cella, U.; Roncella, R.; Santise, M. Testing accuracy and repeatability of UAV blocks oriented with gnss-supported aerial triangulation. *Remote Sens.* **2017**, *9*, 172. [CrossRef]
63. Stöcker, C.; Nex, F.; Koeva, M.; Gerke, M. Quality assessment of combined IMU/GNSS data for direct georeferencing in the context of UAV-based mapping. In Proceedings of the International Archives Photogrammetry Remote Sensing and Spatial Information Sciences (ISPRS Archives), Bonn, Germany, 4–7 September 2017. [CrossRef]
64. Jurjević, L.; Gašparović, M.; Milas, A.S.; Balenović, I. Impact of UAS image orientation on accuracy of forest inventory attributes. *Remote Sens* **2020**, *12*, 404. [CrossRef]
65. Tisdale, J.; Kim, Z.; Hedrick, J. Autonomous UAV path planning and estimation. *IEEE Robot. Autom. Mag.* **2009**, *16*, 35–42. [CrossRef]
66. Nex, F.; Remondino, F. UAV for 3D mapping applications: A review. *Appl. Geomat.* **2014**, *6*, 1–15. [CrossRef]
67. Navia, J.; Mondragon, I.; Patino, D.; Colorado, J. Multispectral mapping in agriculture: Terrain mosaic using an autonomous quadcopter UAV. In Proceedings of the International Conference Unmanned Aircraft System (ICUAS), Arlington, VA, USA, 7–10 June 2016. [CrossRef]
68. Luhmann, T.; Fraser, C.; Maas, H.G. Sensor modelling and camera calibration for close-range photogrammetry. *ISPRS J. Photogramm. Remote Sens.* **2016**, *115*. [CrossRef]
69. Kordić, B.; Gašparović, M.; Oberiter, B.L.; Đapo, A.; Vlastelica, G. Spatial data performance test of mid-cost UAS with direct georeferencing. *Period Polytech. Civ. Eng.* **2020**, *52*. [CrossRef]
70. Kaempchen, N.; Dietmayer, K. Data synchronization strategies for multi-sensor fusion. In Proceedings of the IEEE Conference of Intelligent Transportation System, Shanghai, China, 12–15 October 2003.
71. Gašparović, M. Radiometric Equalization of Textures on the Photorealistic 3D Models. Ph.D. Thesis, Faculty of Geodesy, Zagreb, Croatia, May 2015.

72. Burggraaff, O.; Schmidt, N.; Zamorano, J.; Pauly, K.; Pascual, S.; Tapia, C.; Spyrakos, E.; Snik, F. Standardized spectral and radiometric calibration of consumer cameras. *Opt. Express* **2019**, *14*, 19075–19101. [CrossRef]

73. Kim, S.J.; Pollefeys, M. Robust radiometric calibration and vignetting correction. *IEEE Trans. Pattern. Anal. Mach. Intell.* **2008**, *30*, 562–576. [CrossRef]

74. Sinha, S.N.; Pollefeys, M. Synchronization and calibration of camera networks from silhouettes. In Proceedings of the 17th International Conference on Pattern Recognition, Cambridge UK, 23-26 August 2004. [CrossRef]

75. Huck, T.; Westenberger, A.; Fritzsche, M.; Schwarz, T.; Dietmayer, K. Precise timestamping and temporal synchronization in multi-sensor fusion. In Proceedings of the IEEE Intelligent Vehicles Symposium (IV), Baden-Baden, Germany, 5–9 June 2011. [CrossRef]

76. Rischbeck, P.; Elsayed, S.; Mistele, B.; Barmeier, G.; Heil, K.; Schmidhalter, U. Data fusion of spectral, thermal and canopy height parameters for improved yield prediction of drought stressed spring barley. *Eur. J. Agron.* **2016**, *78*, 44–59. [CrossRef]

77. Liao, W.; Huang, X.; Van Coillie, F.; Gautama, S.; Pižurica, A.; Philips, W.; Liu, H.; Zhu, T.; Shimoni, M.; Moser, G.; et al. Processing of multiresolution thermal hyperspectral and digital color data: Outcome of the 2014 IEEE GRSS data fusion contest. *IEEE J. Sel. Top Appl. Earth Obs. Remote Sens.* **2015**, 2984–2996. [CrossRef]

78. Remondino, F.; Gruen, A. Multi-sensor 3D documentation of the Maya site of Copan. In Proceedings of the 22nd CIPA Symposium, Kyoto, Japan, 11–15 October 2009.

79. Mathews, A.J.; Jensen, J.L.R. Visualizing and quantifying vineyard canopy LAI using an unmanned aerial vehicle (UAV) collected high density structure from motion point cloud. *Remote Sens.* **2013**, *5*, 2164–2183. [CrossRef]

80. Jay, S.; Rabatel, G.; Hadoux, X.; Moura, D.; Gorretta, N. In-field crop row phenotyping from 3D modeling performed using Structure from Motion. *Comput. Electron. Agric.* **2015**, *110*, 70–77. [CrossRef]

81. Lucieer, A.; Jong, S.M.d.; Turner, D. Mapping landslide displacements using Structure from Motion (SfM) and image correlation of multi-temporal UAV photography. *Prog. Phys. Geogr.* **2014**, *38*, 97–116. [CrossRef]

82. Sarakinou, I.; Papadimitriou, K.; Georgoula, O.; Patias, P. Underwater 3D modeling: Image enhancement and point cloud filtering. *Int. Arch. Photogramm. Remote Sens. Spat. Inf. Sci.–ISPRS Arch.* **2016**, *41*. [CrossRef]

83. Zhang, Z.; Gerke, M.; Vosselman, G.; Yang, M.Y. Filtering photogrammetric point clouds using standard lidar filters towards dtm generation. *ISPRS Ann. Photogramm. Remote Sens. Spat. Inf. Sci.* **2018**, *2*. [CrossRef]

84. Zeybek, M.; Şanlıoğlu, İ. Point cloud filtering on UAV based point cloud. *Meas. J. Int. Meas. Confed.* **2019**, *133*, 99–111. [CrossRef]

85. Pintus, R.; Gobbetti, E.; Callieri, M.; Dellepiane, M. *Techniques for Seamless Color Registration and Mapping on Dense 3D Models*, 1st ed.; Springer: Berlin/Heidelberg, Germany, 2017; pp. 355–376. [CrossRef]

86. Dvorožňák, M.; Nejad, S.S.; Jamriška, O.; Jacobson, A.; Kavan, L.; Sýkora, D. Seamless Reconstruction of Part-Based High-Relief Models from Hand-Drawn Images. Available online: https://www.cs.utah.edu/~{}ladislav/dvoroznak18seamless/dvoroznak18seamless.pdf (accessed on 17 August 2018).

87. Duro, D.C.; Franklin, S.E.; Dubé, M.G. A comparison of pixel-based and object-based image analysis with selected machine learning algorithms for the classification of agricultural landscapes using SPOT-5 HRG imagery. *Remote Sens. Environ.* **2012**, *118*, 259–272. [CrossRef]

88. Tzotsos, A. A support vector machine approach for object based image. In Proceedings of the 1st International Conference Object-Based Image Analysis, Salzburg, Austria, 4–5 July 2006.

89. Gašparović, M.; Zrinjski, M.; Barković, Đ.; Radočaj, D. An automatic method for weed mapping in oat fields based on UAV imagery. *Comput. Electron. Agric.* **2020**, *173*, 105385. [CrossRef]

90. Peña, J.M.; Gutiérrez, P.A.; Hervás-Martínez, C.; Six, J.; Plant, R.E.; López-Granados, F. Object-based image classification of summer crops with machine learning methods. *Remote Sens.* **2014**, *6*, 5019–5041. [CrossRef]

91. Gašparović, M.; Zrinjski, M.; Gudelj, M. Automatic cost-effective method for land cover classification (ALCC). *Comput. Environ. Urban Syst.* **2019**, *76*, 1–10. [CrossRef]

92. Jawak, S.D.; Luis, A.J. A spectral index ratio-based Antarctic land-cover mapping using hyperspatial 8-band WorldView-2 imagery. *Polar Sci.* **2013**, *7*, 18–38. [CrossRef]

93. Morsy, S.; Shaker, A.; El-Rabbany, A.; Larocque, P.E. Airborne multispectral lidar data for land-cover classification and land/water mapping using different spectral indexes. *ISPRS Ann. Photogramm. Remote Sens. Spat. Inf. Sci.* **2016**, *3*. [CrossRef]

94. Hall-Beyer, M. Practical guidelines for choosing GLCM textures to use in landscape classification tasks over a range of moderate spatial scales. *Int. J. Remote Sens.* **2017**, *38*, 1312–1338. [CrossRef]

95. Kuffer, M.; Pfeffer, K.; Sliuzas, R.; Baud, I. Extraction of slum areas from VHR imagery using GLCM variance. *IEEE J. Sel. Top. Appl. Earth. Obs. Remote Sens.* **2016**, *9*, 1830–1840. [CrossRef]

96. Raiffa, H.; Schlaifer, R. *Applied Statistical Decision Theory*, 1st ed.; Harvard University Press: Boston, MA, USA, 1961.

97. Haralick, R.M.; Dinstein, I.; Shanmugam, K. Textural features for image classification. *IEEE Trans. Syst. Man. Cybern.* **1973**, *2*, 109–130. [CrossRef]

98. Gašparović, M.; Dobrinić, D. Comparative assessment of machine learning methods for urban vegetation mapping using multitemporal sentinel-1 imagery. *Remote Sens.* **2020**, *12*, 1952. [CrossRef]

99. Gavarini, C. Seismic risk in historical centers. *Soil Dyn. Earthq. Eng.* **2001**, *21*, 459–466. [CrossRef]

100. Papathoma-Köhle, M. Vulnerability curves vs. Vulnerability indicators: Application of an indicator-based methodology for debris-flow hazards. *Nat. Hazards Earth. Syst. Sci.* **2016**, *18*. [CrossRef]

101. Goretti, A.; Di Pasquale, G. An overview of post-earthquake damage assessment in Italy. In Proceedings of the EERI Invitational Workshop. An Action to Develop Earthquake Damage and Loss Data Protocols, Pasadena, CA, USA, 19–20 September 2002.

© 2020 by the authors. Licensee MDPI, Basel, Switzerland. This article is an open access article distributed under the terms and conditions of the Creative Commons Attribution (CC BY) license (http://creativecommons.org/licenses/by/4.0/).

Article

Mode Shape-Based Damage Detection Method (MSDI): Experimental Validation

Ivan Duvnjak *, Domagoj Damjanović, Marko Bartolac and Ana Skender

Faculty of Civil Engineering, University of Zagreb, 10000 Zagreb, Croatia; domagoj.damjanovic@grad.unizg.hr (D.D.); marko.bartolac@grad.unizg.hr (M.B.); ana.skender@grad.unizg.hr (A.S.)
* Correspondence: ivan.duvnjak@grad.unizg.hr

Abstract: The main principle of vibration-based damage detection in structures is to interpret the changes in dynamic properties of the structure as indicators of damage. In this study, the mode shape damage index (MSDI) method was used to identify discrete damages in plate-like structures. This damage index is based on the difference between modified modal displacements in the undamaged and damaged state of the structure. In order to assess the advantages and limitations of the proposed algorithm, we performed experimental modal analysis on a reinforced concrete (RC) plate under 10 different damage cases. The MSDI values were calculated through considering single and/or multiple damage locations, different levels of damage, and boundary conditions. The experimental results confirmed that the MSDI method can be used to detect the existence of damage, identify single and/or multiple damage locations, and estimate damage severity in the case of single discrete damage.

Keywords: damage detection; plate-like structure; operational modal analysis; mode shape; modal assurance criterion (MAC) matrix; mode shape damage index (MSDI); discrete damages

Citation: Duvnjak, I.; Damjanović, D.; Bartolac, M.; Skender, A. Mode Shape-Based Damage Detection Method (MSDI): Experimental Validation. *Appl. Sci.* **2021**, *11*, 4589. https://doi.org/10.3390/app11104589

Academic Editor: César M. A. Vasques

Received: 19 April 2021
Accepted: 15 May 2021
Published: 18 May 2021

Publisher's Note: MDPI stays neutral with regard to jurisdictional claims in published maps and institutional affiliations.

Copyright: © 2021 by the authors. Licensee MDPI, Basel, Switzerland. This article is an open access article distributed under the terms and conditions of the Creative Commons Attribution (CC BY) license (https://creativecommons.org/licenses/by/4.0/).

1. Introduction

During service life, structures are subjected to different loads, external factors, and unpredictable influences that can cause considerable structural damage. Regular inspections of structures are essential for early damage detection, analysis, and repair of damaged structures in order to extend the service life while assuring safety and reliability. Due to the lack of regular inspections and poor maintenance of structures, the damage can decrease the bearing capacity and endanger safety. The widely accepted structure inspection is based on visual inspections or locally limited nondestructive testing methods for damage assessment such as acoustic or ultrasonic methods, magnet field methods, radiography, eddy current methods, thermal field methods, and X-rays [1]. The mentioned methods have numerous limitations, such as the small inspection area and the location of damage needing to be known in advance. The problem is that the part of the structure where the inspection is carried out has to be accessible.

The difficulties in damage detection can be overcome by analyzing the dynamic response of structures obtained using the vibration-based monitoring system [2]. Global behavior of the structure is defined by its dynamic properties that can be determined using relatively simple measurement methods at a single or several locations in the structure. The main reason for using vibration-based monitoring is that structural damages such as changes in boundary conditions and bending cracks cause the loss of stiffness and consequently a change in dynamic properties (natural frequencies, mode shapes, and damping ratios).

Four levels of damage detection in structures are defined as follows [3]: detecting the presence of damage, detecting the damage location, determining the severity of the damage, and predicting the remaining service life. Different methods have been proposed on the basis of the change in natural frequencies [4–9]; mode shapes [10,11]; and their derivatives such as the change in modal flexibility [12,13], the change in modal curvature [14,15], the change in modal strain energy [16–18], and other modal methods used for damage detection on plate-like structures [19]. Furthermore, a number of non-modal methods are used for damage detection on plate-like structures, such as the frequency response function (FRF) method [20] and guided waves approach [21]. They are mainly used with some other algorithms (artificial neural networks, genetic algorithm, Bayesian approach, etc.), requiring additional effort to detect damage.

Mentioned methods are primarily used for the first two levels of damage assessment, i.e., damage detection and localization. Some of them can even be used to quantify the severity of the damage but mostly to identify the difference between lower and higher levels of damage [7,16,18]. When multiple structural damages are considered, determining damage severity becomes a challenge [22]. Most of the methods are applied to numerical and experimental models of rods, plates [23], and beams with simulated damages or real-life structures such as bridges [10,24].

Modal assurance criterion (MAC) [25–27], among other indicators, is used for damage detection based on mode shape changes. The MAC is a scalar quantity related to the degree of consistency between two mode shapes, and it can be used for the direct comparison of mode shapes in the undamaged and damaged condition. This criterion is not sensitive to small differences between mode shapes and cannot be used for the damage localization, but it can serve as an indicator of damage in the structure [28,29]. In order to detect and localize damages in structures, researchers proposed the mode shape damage index (MSDI) algorithm on the basis of a modified MAC matrix (ΔMAC), which considers mode shapes that are almost identical and excludes dissimilar mode shapes [30]. During this research, finite element analysis (FEA) was performed to obtain the mode shapes of plate models with different damage cases and boundary conditions. It was concluded that the MSDI method can be used to accurately locate single and multiple damages on plate-like structures and that it has the capability to distinguish damages with different levels of severity. By comparing the MSDI method to the l_1-norm regularized finite element model updating method, researchers obtained similar results in damage localization [31].

The advantage of the MSDI method is that it does not require an additional tool, for example, a finite element model or algorithm-based massive measurement data such as artificial intelligence [32], machine learning [33], and deep learning [34] for damage detection. Unlike existing damage detection methods based on mode shape changes, this method is based on a minor difference between compared undamaged and damaged mode shapes. In this way, it is possible to determine the damage at the earliest stage of its occurrence, and at the same time exclude the influence of uncertainties (noise) that may significantly affect the change in mode shapes.

The aim of this study was to validate the MSDI method on the basis of the experimental model of a reinforced concrete (RC) plate under 10 different damage cases. Operational modal analysis (OMA) [35] was used for the determination of dynamic properties (natural frequencies, mode shapes, and damping ratios) of the plate in the undamaged and damaged state. The MSDI values were calculated for all damage cases and graphically presented.

The article is structured as follows. The governing equations of the MSDI algorithm are shown in Section 2.1. The experimental setup and procedure as well as the damage simulation are presented in Section 2.2. The results are summarized and discussed in Section 3. Section 3.1 deals with the effect of the damage level on the MSDI values at a single damage location. Section 3.2 deals with the effect of the damage zone size, and Section 3.3 deals with damage detection at multiple damage locations. Section 4 presents the conclusions.

2. Materials and Methods

2.1. The Mode Shape Damage Index (MSDI)

The MAC criterion provides a degree of consistency between two states (undamaged and damaged), and it ranges between 0 and 1. The MAC matrix is defined as

$$
MAC_{(k,l)} = \frac{\left| \{\phi_k^u\}^T \{\phi_l^d\} \right|^2}{\left(\{\phi_k^u\}^T \{\phi_k^u\} \right)\left(\{\phi_l^d\}^T \{\phi_l^d\} \right)},
\tag{1}
$$

where $\{\phi_k^u\}$ represents the k-th mass-normalized mode shape vector in the undamaged state and $\{\phi_l^d\}$ represents the l-th mass-normalized mode shape vector in the damaged state. The values of the diagonal elements of the MAC matrix represent the correlation between corresponding mode shapes. If there is no damage and no noise in the signal, the diagonal elements of the MAC matrix are equal to 1, while values that are less than 1 reveal a weak correlation between the two mode shapes that can indicate the presence of damage (reduction of stiffness). The non-diagonal elements of the MAC matrix are usually equal to 0 because their values represent the correlation of inconsistent modes. Therefore, the MSDI damage localization algorithm uses only the diagonal elements ($k = l$) of the MAC matrix represented by the squared value of the trace of the MAC matrix defined as

$$
\gamma_{trMAC} = (trMAC)^2.
\tag{2}
$$

This value ranges between zero and n^2, where n represents the number of correlated mode shapes. If $\gamma_{trMAC} = n^2$, then the compared mode shape vectors are highly correlated and almost identical, which means that there is no change in stiffness. When the value tends to 0, the vectors of the compared mode shapes are entirely inconsistent. The uncertainties such as noise in the signal, which can occur in the experimental determination of mode shapes, are ignored.

Further, a modified matrix named ΔMAC was developed, which is sensitive even to a small difference between correlated mode shapes in the undamaged and damaged state. The diagonal elements of the ΔMAC matrix are defined as follows

$$
\Delta\alpha_{kl} = \alpha_{kl}{}^{\gamma_{trMAC}}, \; k = l,
\tag{3}
$$

where α_{kl} and $\Delta\alpha_{kl}$ are the diagonal elements of the original MAC matrix and the modified ΔMAC matrix, respectively. Theoretically, if there is no change in stiffness and no noise in the signal, the mode shapes match entirely, and the modified ΔMAC matrix equals the original MAC matrix.

The mode shape damage index (MSDI) is a damage indicator based on the difference between modified modal displacements in the undamaged and damaged state in each i-th node of a structural element (i.e., plate) as follows

$$
|MSDI_i| = \frac{\left(\overline{\Phi}_k^u\right)_i - \left(\overline{\Phi}_l^d\right)_i}{\left(\overline{\Phi}_k^u\right)_i},
\tag{4}
$$

where $\left(\overline{\Phi}_k^u\right)_i$ and $\left(\overline{\Phi}_l^d\right)_i$ are the modified modal displacements in each i-th node given by the following equations

$$
\left(\overline{\Phi}_k^u\right)_i = \sum_{k=1}^{n} \left((\phi_k^u)_i^2 \cdot \Delta\alpha_{kl} \right),
\tag{5}
$$

$$
\left(\overline{\Phi}_l^d\right)_i = \sum_{l=1}^{n} \left(\left(\phi_l^d\right)_i^2 \cdot \Delta\alpha_{kl} \right),
\tag{6}
$$

where $\left(\phi_k^u\right)_i$ and $\left(\phi_l^d\right)_i$ denote the values of modal displacements of the k-th mode shape in the undamaged state and the l-th mode shape in the damaged state in each i-th node of the plate. The normalized modal displacements are squared to highlight a local change of the mode shape in the damaged state compared to the undamaged state. Furthermore, modal displacements are multiplied by the corresponding value $\Delta\alpha_{kl}$ calculated from the modified matrix ΔMAC. In this way, only mode shapes with a reasonable degree of correlation are considered, and the ones that do not match are excluded from the damage localization. If even small damages are present at the i-th node, the MSDI value becomes a nondimensional negative value, presented as an absolute value. Theoretically, if there is no damage, i.e., no changes in the stiffness of the system (assuming there is no change in mass and damping), the MSDI value equals 0 in each i-th node of the plate.

2.2. Dynamic Testing of the Plate Model

To further validate the MSDI method and investigate its robustness, we performed experimental research on a model of a reinforced concrete (RC) plate with plan dimensions 2.3 × 1.55 m and 7 cm of thickness. The plate was reinforced with the Q196 reinforcement mesh in the tension zone, and the thickness of the concrete cover was 2.5 cm. The mechanical properties of the concrete, i.e., the characteristic compressive strength (f_{ck}) and the modulus of elasticity (E), were determined on 28-day-old test specimens according to standards EN 12390-3 and EN 12390-13, respectively. The experimentally obtained values were f_{ck} = 70 MPa and E = 36 GPa. The model of an RC plate was divided into rectangular elements (16 cm × 15 cm); rows were marked in letters (A–J) and columns in numbers (1–14), as shown in Figure 1. A total of 165 (1–165) measurement points on the RC plate were used for the experimental modal analysis.

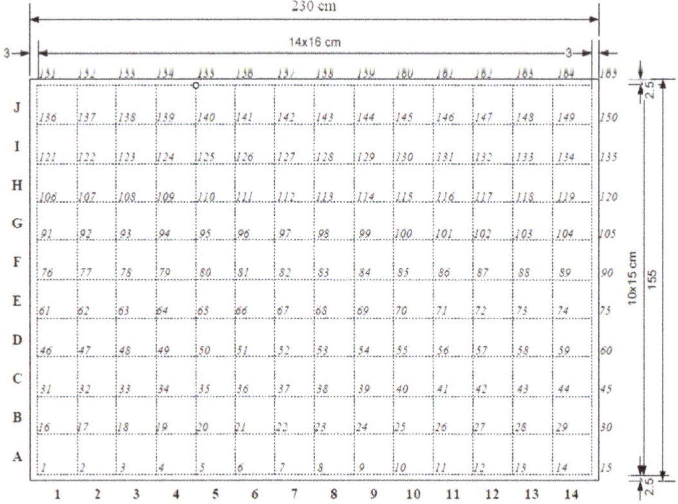

Figure 1. Measurement points on the RC plate.

The operational modal analysis (OMA) [35] was applied for the detection of dynamic properties (natural frequencies, mode shapes, and damping ratios). The measurements were taken with the Brüell&Kjaer Multi-Analyzer system (type 3560-C) using five channels. The modal analysis, including signal processing and modal extraction procedures, was further performed using Pulse software. Five piezoelectric accelerometers (Brüell&Kjaer 4508-B, sensitivity 10 mV/ms^{-2}) were used for acceleration measurements. OMA was conducted by moving four accelerometers through 41 measuring stages using one referent measurement point at position no. 155 (41 × 4 + 1 = 165 measurement points) (Figure 1).

The impact hammer was used to excite the plate randomly. This kind of excitation aims to simulate an operational environment where the excitation is difficult or impossible to measure. Methods of frequency domain decomposition (FDD) [36] and enhanced frequency domain decomposition (EFDD) [37] were used for the estimation of modal parameters. The procedure is based on the singular value decomposition (SVD) of the power spectral density (PSD) matrix of the measured responses.

To achieve constant boundary conditions, we suspended the plate using very elastic ropes (Figure 2). The applied boundary conditions, a nearly ideal free-free set-up, were quickly implemented and easily repeated in the laboratory after each testing phase. The plate was tested in an undamaged state and then discrete damages were applied successively by removing parts of the surface from the top of the concrete plate.

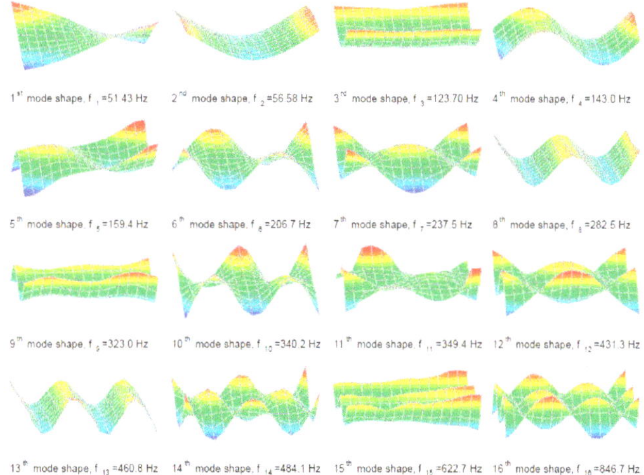

Figure 2. Suspended plate with discrete damages.

The plate was again tested after each successive damage simulation. A total of 16 experimentally obtained vertical mode shapes for the undamaged state (Figure 3) and each phase of the damaged state were observed in the analysis.

Figure 3. Sixteen mode shapes obtained experimentally on the undamaged plate.

Simulation of Discrete Damages

In the experimental research, discrete damage was applied by removing a part of the surface from the top of the concrete plate (Figure 4). The size of the discrete damaged rectangle was 15 × 16 cm, and the depth varied from 2 to 3 cm. The advantage of discrete damages lies in the fact that the damage location and the change in stiffness are precisely defined. Furthermore, it is possible to simulate a full range of damages, and the MSDI method can be tested using more than one damage location. In sequence, a total of 10 discrete damage cases were simulated on the concrete plate. It is known from the literature that stiffness has a dominant influence on the changes in mode shapes [38]. Therefore, the loss of mass was neglected in the experimental part of this research. Furthermore, the method was previously validated on numerical models [30,31] based only on the reduction of stiffness without including the effect of changes in mass.

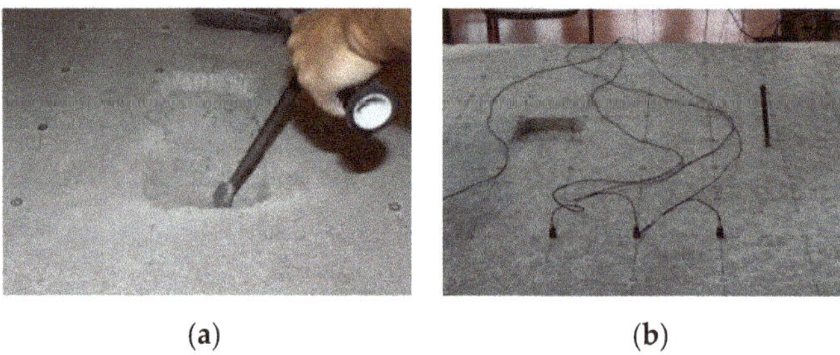

(**a**) (**b**)

Figure 4. Discrete damages on the RC plate: (**a**) chiseling; (**b**) one of the damage locations on the concrete plate.

The first discrete damage was applied at the element E8 (damage case DC1) by reducing the concrete thickness for 2 cm. Afterward, an additional 1 cm of concrete was removed (damage case DC2). The remaining discrete damages were simulated by removing 3 cm of concrete at the plate elements as follows: F8, E7, C4, C3, J9, A12, H13, and H4. Locations of all discrete damages are presented in Figure 5. Different damage cases are listed in Table 1.

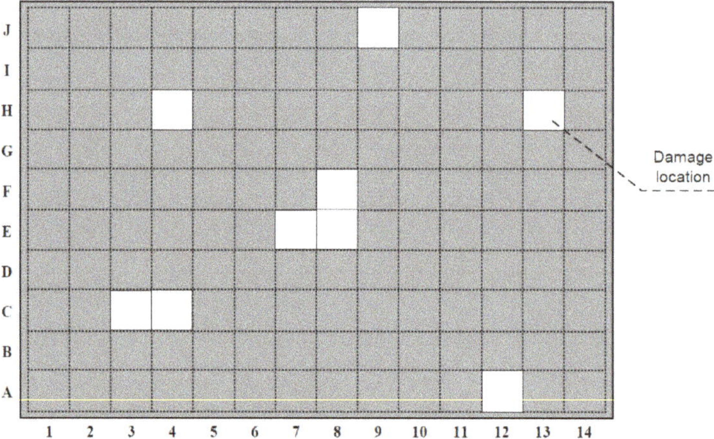

Figure 5. Discrete damage locations on the concrete plate.

Table 1. Damage cases simulated on the concrete plate.

Damage Case	Elements	Depth (cm)
DC1	E8	2
DC2	E8	3
DC3	E8 + F8	3
DC4	DC3 + E7	3
DC5	DC4 + C4	3
DC6	DC5 + C3	3
DC7	DC6 + J9	3
DC8	DC7 + A12	3
DC9	DC8 + H13	3
DC10	DC9 + H4	3

3. Results and Discussion

On the basis of experimental research, we performed a sensitivity analysis of the MSDI method, considering the effect of the damage level, the effect of the damage zone size, and ultimately the effect of multiple damages.

3.1. Effect of Damage Level at a Single Damage Location

The first two damage cases at a single damage location were analyzed to determine the effect of the damage level when using the MSDI method. Damage cases DC1 and DC2 at element E8 simulated two levels of damage, "small" (depth of 2 cm) and "moderate" damage (depth of 3 cm). The damage location was determined by the MSDI method and presented as a colored graphical representation in Figure 6 for the first two damage cases. The MSDI values were calculated according to Equation (4). The peak values indicate the location of the damage on the plate in 3D view, and the damaged area is presented in 2D view. The exact location of the discrete damage, the element E8, was accurately identified. Moreover, the increase from small to moderate damage was followed by the increase of the MSDI value.

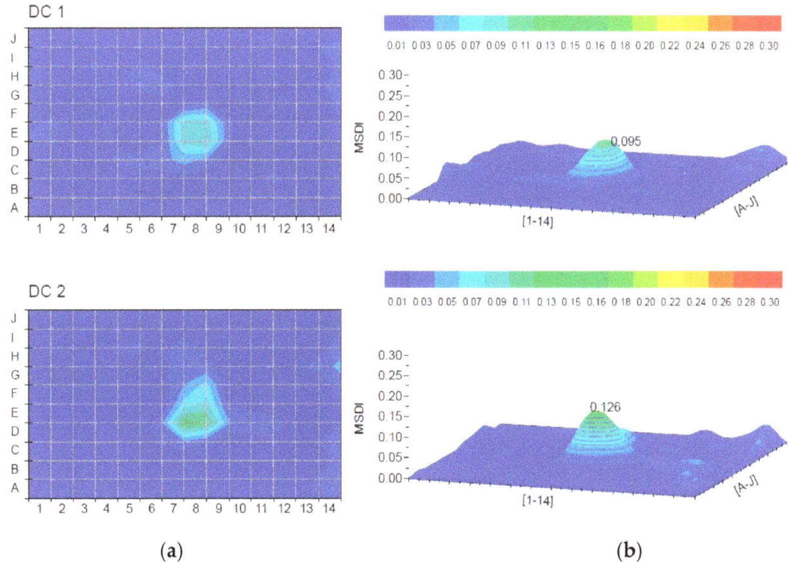

(a)   (b)

Figure 6. Damage localization for damage cases DC1 and DC2: (**a**) 2D; (**b**) peak values in 3D.

Furthermore, it is interesting to compare diagonal elements of the MAC and ΔMAC matrices both in numerical and experimental examples. In the numerical analysis, almost all coefficients participate with the maximum value 1 when damage is detected [30] because there are no measurement uncertainties. The comparison of experimentally obtained values is presented in Figure 7 for the first 16 mode shapes of the plate for damage cases DC1 and DC2. Some of the mode shapes for both damage cases were inconsistent (Figure 7a, mode shapes 11, 13, 14; Figure 7b, mode shapes 11, 13, 14, 15) in the undamaged and damaged states as a result of noise in the measurement signal and some other uncertainties (experimentally unrepeatable boundary conditions). It is shown that only a few mode shapes participated with values close to 1 (Figure 7a, mode shapes 1, 4, 6), whose weighting factor $\Delta\alpha_{kl}$ on modal displacements (Equations (5) and (6)) was sufficient for damage localization. Mentioned uncertainties are challenging to simulate or predict by numerical analysis. Thus, it is essential to emphasize the importance of validating damage detection methods on the basis of experimental results.

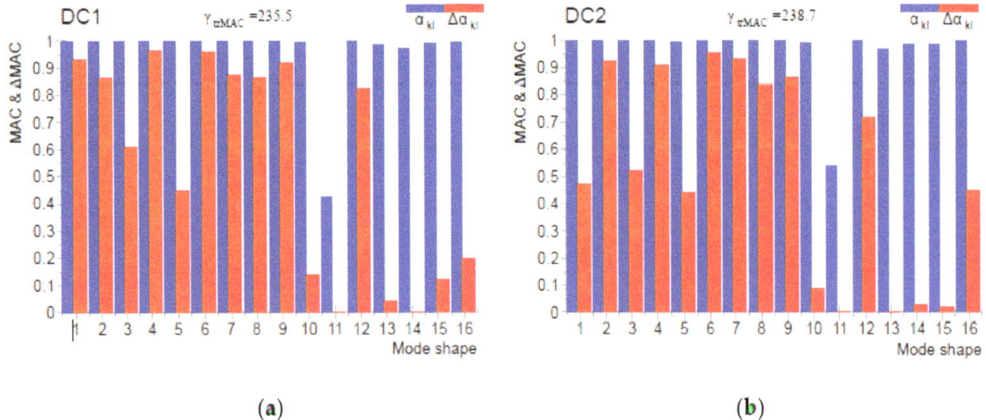

Figure 7. Diagonal elements of the MAC and ΔMAC matrices: (**a**) damage case DC1; (**b**) damage case DC2.

3.2. Effect of Damage Zone Size

In this section, the influence of the damage zone size on the sensitivity of the MSDI method was considered. The third level of damage can be described as "large" damage, simulated by the loss of depth which is equal to "moderate" damage (3 cm) but with more than one damaged element. These substantial damages were simulated as damage cases DC3 (elements E8 and F8) and DC4 (elements E8, F8, and E7). Figure 8 summarizes the results of the effect of the damage zone size. In both damage cases, the damage was accurately localized. Moreover, the method can be used to determine larger damaged zones, i.e., several adjacent damaged elements. The MSDI values kept increasing with the addition of more damage.

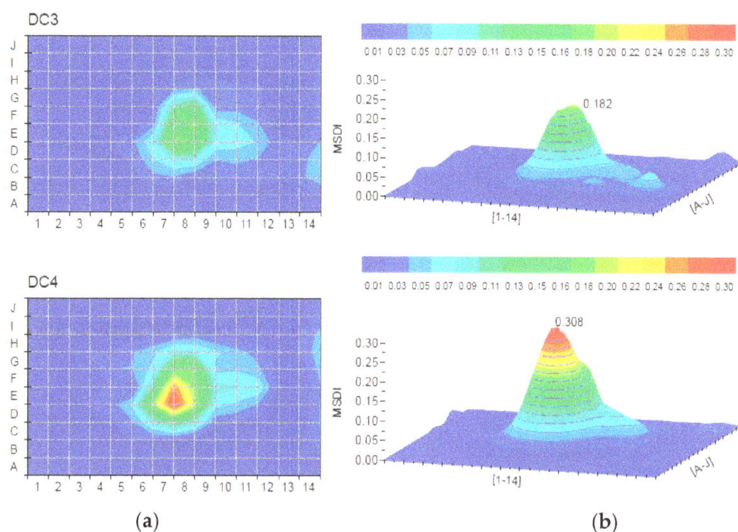

Figure 8. Damage localization for damage cases DC3 and DC4: (**a**) 2D; (**b**) peak values in 3D.

3.3. Effect of Multiple Damages

The remaining damage cases were analyzed to primarily investigate the effect of multiple damages on the MSDI method. In damage cases DC5 (DC4 + element C4) and DC6 (DC5 + element C3), the effect of multiple damages and the effect of the damage zone size were considered by adding another damage location on the plate. When we compared the results for both damage cases in Figure 9, we found that the MSDI method was able to register both damages, regardless of the damage zone size. The peak MSDI values at the two locations coincided with the reduction of stiffness. Moreover, the damage severity was smaller at the single damaged element C4 (see Figure 9, DC5) than the damaged zone C4-C3 (see Figure 9, DC6).

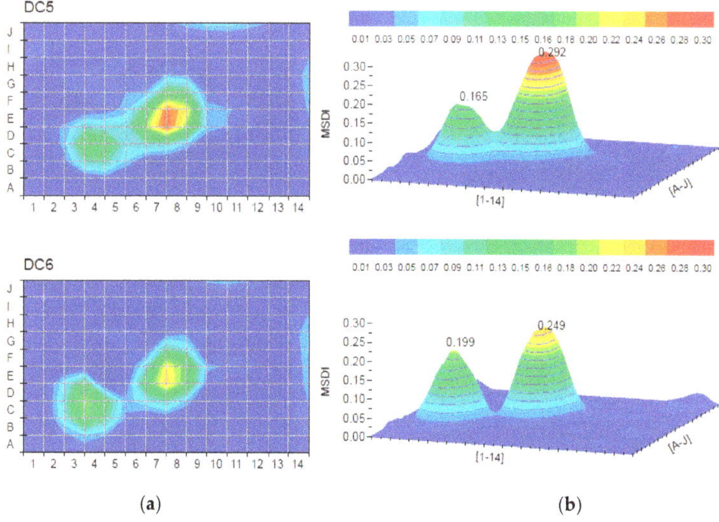

Figure 9. Damage localization for damage cases DC5 and DC6: (**a**) 2D; (**b**) peak values in 3D.

The following two multiple damage cases DC7 and DC8 included damaged elements at the free edges of the plate. The results shown in Figure 10 indicate that the method effectively identified multiple damages regardless of the location of the damaged zone. Nevertheless, as previously concluded in the finite element analysis [30], the damage severity was influenced by the damage location. Therefore, the severity of multiple damages cannot be determined by using this method.

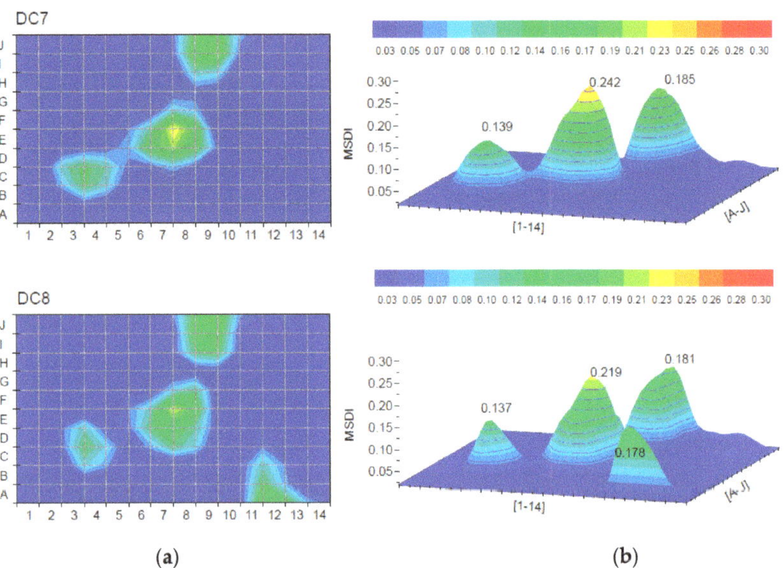

(a)   (b)

Figure 10. Damage localization for damage cases DC7 and DC8: (**a**) 2D; (**b**) peak values in 3D.

The last two damage cases DC9 and DC10 considered two additional random single damage locations in elements H13 and H4. The results of simulated damages are shown in Figure 11, and it can be found that all simulated damages were accurately localized by the proposed algorithm regardless of the damaged zone size or the damage location.

The diagonal elements of the MAC and ΔMAC matrices are shown for the damage cases DC5 to DC10 in Figure 12. It is noted that the number of mode shapes that were participating in damage detection decreased with the introduction of new damages. The first reason was that the intensity (α_{kl} value) decreased due to the low degree of consistency between estimated modal vectors in the undamaged and damaged states. The second reason was that some mode shapes entirely disappeared due to a significant change of consistency between estimated modal vectors. Even though a smaller number of mode shapes were used for damage detection based on the MSDI method, they were still able to provide useful information about the location of structural damage.

The accuracy of the MSDI method for damage localization depends on the number of measurement points (degrees of freedom) for some element or structure. In order to apply this method in practice within the structural health monitoring system (e.g., bridges, tall buildings), one must optimize the number of measuring points.

To obtain better reliability of the damage detection by using the presented method, we recommend as many experimentally obtained mode shapes as possible. In the paper, all experimentally observed mode shapes are presented and were used in the MSDI method, although a large number of mode shapes (for instance, Figure 12, DC 9) did not participate at all in damage detection or participate with significantly reduced impact.

The environmental (e.g., temperature) and serviceability effects were not considered in this research. It is necessary to consider the effects of temperature changes depending on the boundary conditions of the structure. Under the influence of different temperatures, the mode shapes can be changed significantly, and this can affect the MSDI analysis.

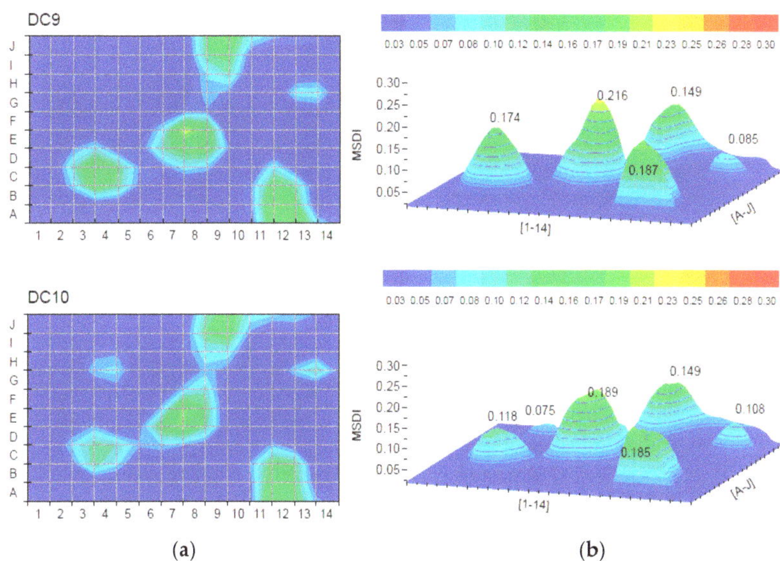

Figure 11. Damage localization for damage cases DC9 and DC10: (**a**) 2D; (**b**) peak values in 3D.

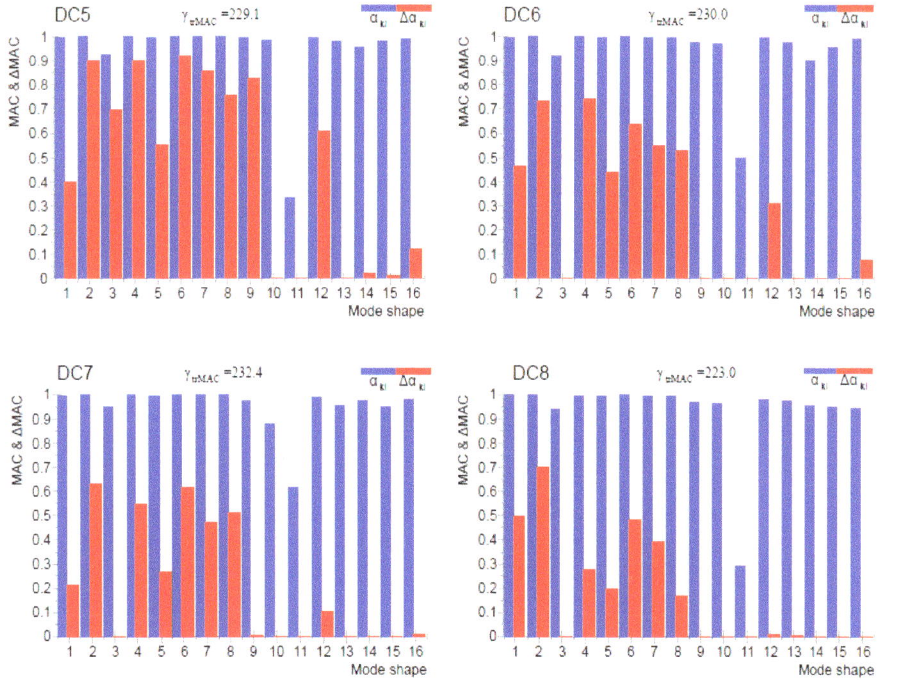

Figure 12. *Cont.*

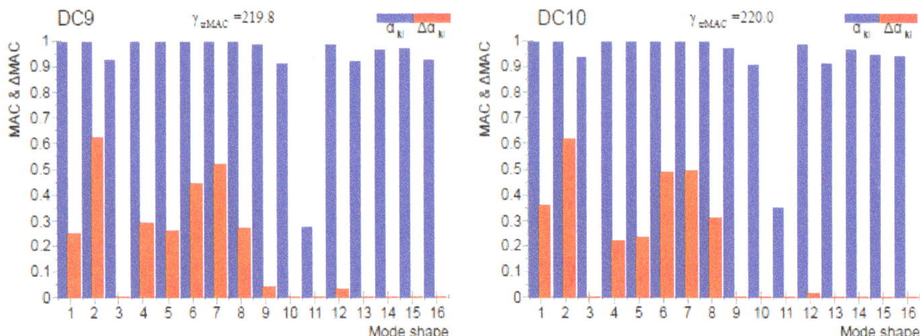

Figure 12. Diagonal elements of the MAC and ΔMAC matrices for damage cases DC5–DC10.

4. Conclusions

In this study, experimental validation of the mode shape damage index (MSDI) method for damage detection was performed on a reinforced concrete plate. Ten discrete damage cases were simulated by removing a part of the top layer on the concrete plate. On the basis of the measurement of the vibration response, we determined the dynamic properties of the plate in the undamaged and damaged conditions by applying the operational modal analysis. This method is particularly challenging due to many factors that can affect the accuracy of measurement and, therefore, can directly affect the damage analysis results. Namely, damage-induced changes in mode shapes are minimal, and every error in the experimental analysis of mode shapes can significantly affect the accuracy of damage detection. By applying the MSDI method on the experimental model, it is evident that the mode shapes that did not match for any reason were excluded from the damage analysis. Even though a smaller number of mode shapes were used for damage detection in the MSDI method, they can still provide useful information about the location of structural damage.

The conclusions from previous numerical research are confirmed. The MSDI method can be used to detect the existence of damage, identify single and/or multiple damage locations, and distinguish damages with different levels of severity. It must be emphasized that the severity of damage detected with this method depends on boundary conditions and the location of the damage.

The MSDI method can easily be implemented in the vibration-based damage detection system. This method does not require the numerical model for the damage assessment, and the measurement is simplified due to the fact it uses only the ambient response of the structure for the estimation of modal parameters. It is also possible to combine this method with some of the existing methods. For example, the change in natural frequency of structure can serve as an indicator of the damage existence, and then by applying the MSDI method, the damage location can be determined.

Although this method shows promising results and successful validation through experimental tests, there are still many challenges for future research in structural damage identification. By applying the MSDI method, it is still not possible to determine damage severity. Hence, future research is recommended in that direction. Furthermore, various uncertainties that may affect the results by using this method need to be considered, for instance, inconsistent boundary conditions, various ambient conditions (e.g., temperature, humidity), various types of damages (e.g., bending cracks, corrosion, wear), and serviceability effects (traffic) in a laboratory environment and in real structures.

Author Contributions: All authors have contributed to the contents within this paper. Conceptualization, I.D.; methodology, I.D.; experimental research, I.D., software, D.D. and M.B.; validation, I.D. and A.S.; literature review, I.D. and A.S.; writing—original draft preparation, I.D.; writing—review and editing, A.S.; visualization, D.D. and M.B.; supervision, D.D. All authors have read and agreed to the published version of the manuscript.

Funding: This research received no external funding.

Institutional Review Board Statement: Not applicable.

Informed Consent Statement: Not applicable.

Conflicts of Interest: The authors declare no conflict of interest.

References

1. Cornwell, P.; Farrar, C.R.; Doebling, S.W.; Sohn, H. Environmental Variability of Modal Properties. *Exp. Tech.* **1999**, *23*, 45–48. [CrossRef]
2. Wang, F.L.; Chan, T.H.T. Review of Vibration-Based Damage Detection and Condition Assessment of Bridge Structures Using Structural Health Monitoring. In *The Second Infrastructure Theme Postgraduate Conference: Rethinking Sustainable Development: Planning, Engineering, Design and Managing Urban Infrastructure*; Queensland University of Technology: Brisbane, Australia, 2009; pp. 35–47.
3. Rytter, A. Vibrational Based Inspection of Civil Engineering Structures. Ph.D. Thesis, Department of Building Technology and Structural Engineering, Aalborg University, Aalborg, Denmark, 1993.
4. Salawu, O.S. Detection of Structural Damage through Changes in Frequency: A Review. *Eng. Struct.* **1997**, *19*, 718–723. [CrossRef]
5. Messina, A.; Williams, E.J.; Contursi, T. Structural Damage Detection by a Sensitivity and Statistical-Based Method. *J. Sound Vib.* **1998**, *216*, 791–808. [CrossRef]
6. Morassi, A. Identification of a Crack in a Rod Based on Changes in a Pair of Natural Frequencies. *J. Sound Vib.* **2001**, *242*, 577–596. [CrossRef]
7. Kim, J.T.; Stubbs, N. Crack Detection in Beam-Type Structures Using Frequency Data. *J. Sound Vib.* **2003**, *259*, 145–160. [CrossRef]
8. Salehi, M.; Ziaei-Rad, S.; Ghayour, M.; Vaziri-Zanjani, M.A. A Structural Damage Detection Technique Based on Measured Frequency Response Functions. *Contemp. Eng. Sci.* **2010**, *3*, 215–226.
9. Yang, Z.; Chen, X.; Yu, J.; Liu, R.; Liu, Z.; He, Z. A Damage Identification Approach for Plate Structures Based on Frequency Measurements. *Nondestruct. Test. Eval.* **2013**, *28*, 321–341. [CrossRef]
10. Beskhyroun, S.; Wegner, L.D.; Sparling, B.F. New Methodology for the Application of Vibration-Based Damage Detection Techniques. *Struct. Control Health Monit.* **2012**, *19*, 632–649. [CrossRef]
11. Xu, Y.; Zhu, W. Non-Model-Based Damage Identification of Plates Using Measured Mode Shapes. *Struct. Health Monit.* **2017**, *16*, 3–23. [CrossRef]
12. Pandey, A.K.; Biswas, M. Experimental Verification of Flexibility Difference Method for Locating Damage in Structures. *J. Sound Vib.* **1995**, *184*, 311–328. [CrossRef]
13. Catbas, F.N.; Brown, D.L.; Aktan, A.E. Use of Modal Flexibility for Damage Detection and Condition Assessment: Case Studies and Demonstrations on Large Structures. *J. Struct. Eng.* **2006**, *132*, 1699–1712. [CrossRef]
14. Abdel Wahab, M.M.; De Roeck, G. Damage Detection in Bridges Using Modal Curvatures: Application to a Real Damage Scenario. *J. Sound Vib.* **1999**, *226*, 217–235. [CrossRef]
15. Maeck, J.; De Roeck, G. Damage Detection on a Prestressed Concrete Bridge and RC Beams Using Dynamic System Identification. *Key Eng. Mater.* **1999**, *167*, 320–327. [CrossRef]
16. Shi, Z.Y.; Law, S.S.; Zhang, L.M. Structural Damage Detection from Modal Strain Energy Change. *J. Eng. Mech.* **2000**, *126*, 1216–1223. [CrossRef]
17. Hu, H.; Wang, B.T.; Lee, C.H.; Su, J.S. Damage Detection of Surface Cracks in Composite Laminates Using Modal Analysis and Strain Energy Method. *Compos. Struct.* **2006**, *74*, 399–405. [CrossRef]
18. Meruane, V.; Lasen, M.; Droguett, E.L.; Ortiz-Bernardin, A. Modal Strain Energy-Based Debonding Assessment of Sandwich Panels Using a Linear Approximation with Maximum Entropy. *Entropy* **2017**, *19*, 619. [CrossRef]
19. Fan, W.; Qiao, P. Vibration-Based Damage Identification Methods: A Review and Comparative Study. *Struct. Health Monit.* **2011**, *10*, 83–111. [CrossRef]
20. Siow, P.Y.; Ong, Z.C.; Khoo, S.Y.; Lim, K.S. Damage Sensitive PCA-FRF Feature in Unsupervised Machine Learning for Damage Detection of Plate-Like Structures. *Int. J. Struct. Stab. Dyn.* **2021**, *21*. [CrossRef]
21. Zhao, M.; Zhou, W.; Huang, Y.; Li, H. Sparse Bayesian Learning Approach for Propagation Distance Recognition and Damage Localization in Plate-like Structures Using Guided Waves. *Struct. Health Monit.* **2021**, *20*, 3–24. [CrossRef]
22. Kahya, V.; Karaca, S.; Okur, F.Y.; Altunışık, A.C.; Aslan, M. Damage Localization in Laminated Composite Beams with Multiple Edge Cracks Based on Vibration Measurements. *Iran. J. Sci. Technol. Trans. Civ. Eng.* **2020**, *45*, 75–87. [CrossRef]
23. Limongelli, M.P. The Surface Interpolation Method for Damage Localization in Plates. *Mech. Syst. Signal Process.* **2019**, *118*, 171–194. [CrossRef]
24. Catbas, F.N.; Gul, M.; Burkett, J.L. Conceptual Damage-Sensitive Features for Structural Health Monitoring: Laboratory and Field Demonstrations. *Mech. Syst. Signal Process.* **2008**, *22*, 1650–1669. [CrossRef]
25. West, M.; West, W. Illustration of the Use of Modal Assurance Criterion to Detect Structural Changes in an Orbiter Test Specimen. *Proc. Air Conf. Aircr. Struct. Integr.* **1984**, *1*, 1–6.
26. Doebling, S.W.; Farrar, C.R.; Prime, M.B. A Summary Review of Vibration-Based Damage Identification Methods. *Shock Vib. Dig.* **1998**, *30*, 91–105. [CrossRef]

27. Allemang, R.J. The Modal Assurance Criterion—Twenty Years of Use and Abuse. In Proceedings of the IMAC-XX: Conference & Exposition on Structural Dynamics, Los Angeles, CA, USA, 4–7 February 2002; p. 9.
28. Pastor, M.; Binda, M.; Harčarik, T. Modal Assurance Criterion. In *Procedia Engineering*; Elsevier: Amsterdam, The Netherlands, 2012; Volume 48, pp. 543–548. [CrossRef]
29. Tatar, A.; Niousha, A.; Rofooei, F.R. Damage Detection in Existing Reinforced Concrete Building Using Forced Vibration Test Based on Mode Shape Data. *J. Civ. Struct. Health Monit.* **2017**, *7*, 123–135. [CrossRef]
30. Duvnjak, I.; Rak, M.; Damjanović, D. A New Method for Structural Damage Detection and Localization Based on Modal Shapes. In Proceedings of the Fifth International Symposium on Life-Cycle Civil Engineering (IALCCE 2016), Delft, The Netherlands, 16–19 October 2016.
31. Duvnjak, I.; Damjanović, D.; Sabourova, N.; Grip, N.; Ohlsson, U.; Elfgren, L.; Tu, Y. Damage Detection in Structures—Examples. In Proceedings of the IABSE Symposium 2019 Guimarães: Towards a Resilient Built Environment—Risk and Asset Management, Guimarães, Portugal, 27–29 March 2019.
32. Sun, L.; Shang, Z.; Xia, Y.; Bhowmick, S.; Nagarajaiah, S. Review of Bridge Structural Health Monitoring Aided by Big Data and Artificial Intelligence: From Condition Assessment to Damage Detection. *J. Struct. Eng.* **2020**, *146*, 04020073. [CrossRef]
33. Hsieh, Y.-A.; Tsai, Y.J. Machine Learning for Crack Detection: Review and Model Performance Comparison. *J. Comput. Civ. Eng.* **2020**, *34*, 04020038. [CrossRef]
34. Dang, H.V.; Raza, M.; Nguyen, T.V.; Bui-Tien, T.; Nguyen, H.X. Deep Learning-Based Detection of Structural Damage Using Time Series Data. *Struct. Infrastruct. Eng.* **2020**. [CrossRef]
35. Zhang, L.; Brincker, R.; Andersen, P. An Overview of Operational Modal Analysis: Major Development and Issues. In Proceedings of the 1st International Operational Modal Analysis Conference (IOMAC 2005), Copenhagen, Denmark, 26–27 April 2005; pp. 149–160.
36. Brincker, R.; Zhang, L.; Andersen, P. Modal Identification of Output-Only Systems Using Frequency Domain Decomposition. *Smart Mater. Struct.* **2001**, *10*, 441–445. [CrossRef]
37. Gade, S.; Møller, N.B.; Herlufsen, H.; Konstantin-Hansen, H. Frequency Domain Techniques for Operational Modal Analysis. In Proceedings of the 1st International Operational Modal Analysis Conference (IOMAC 2005), Copenhagen, Denmark, 26–27 April 2005; p. 231.
38. Khatibi, M.M.; Ashory, M.R.; Malekjafarian, A.; Brincker, R. Mass-Stiffness Change Method for Scaling of Operational Mode Shapes. *Mech. Syst. Signal Process.* **2012**, *26*, 34–59. [CrossRef]

 *applied sciences*

Article

Characteristics of Plane Gate Vibration and Holding Force in Closing Process by Experiments

Yanzhao Wang [1], Guobin Xu [1,*], Wensheng Li [1,2], Fang Liu [1] and Yu Duan [1]

[1] State Key Laboratory of Hydraulic Engineering Simulation and Safety, Tianjin University, Tianjin 300350, China; wangyanzhao27@tju.edu.cn (Y.W.); 2018205322@tju.edu.cn (W.L.); fangliu@tju.edu.cn (F.L.); duanyu09@tju.edu.cn (Y.D.)

[2] Shenzhen Water Planning & Design Insitute Co., Ltd., Shenzhen 518001, China

* Correspondence: xuguob@tju.edu.cn

Received: 11 August 2020; Accepted: 31 August 2020; Published: 3 September 2020

Abstract: A 1:25 scale physical model test was employed to study the plane gate vibration and holding force under the conditions of the fixed gate opening and closing process, respectively. We paid more attention to the characteristics of the gate vibration, holding force and the failure of gate-closing in closing process. The correlation between gate vibration and holding force was further examined. The results show that vertical vibration is weaker than the lateral and horizontal vibrations in fixed gate opening and is stronger than the lateral and vertical vibrations in closing process. Gate vertical vibration is self-excited vibration with a frequency of 7–14 Hz. Besides, crawl vibration in closing process is related to the upstream water depth. The higher the water level is, the earlier the crawl vibration appears. After the crawl stage, plane gate stops motion at a certain distance from the chamber floor and then the failure of gate-closing happens. Finally, gate vibration in three directions is significantly correlated with the holding force. In closing process, holding force has positive correlation with the vertical vibration and has negative correlation with the lateral and horizontal vibrations. In the crawl stage, the average of correlation coefficient in lateral, vertical and horizontal direction is −0.723, 0.733 and −0.664, respectively. Thus, the influence of gate vibration on holding force should be taken into consideration in determining the hoists capacity.

Keywords: plane gate; physical model test; flow-induced vibration; crawl vibration; holding force; correlation

---

1. Introduction

Vibration problems in water conservancy projects (including dams, gates and plants) caused by spillway is a common phenomenon. It can cause serious consequences and threaten the safe and stable operation of water conservancy projects [1–4]. Flow-induced vibration of plane gate is an important subject in water conservancy engineering. Its vibration mechanism and vibration control method are both the focus of research.

There are many excellent opinions in the literature dealing with the basic mechanism of flow-induced vibration of gate. Hardwick [5] and Jongeling [6], through the experiment, pointed out that the main cause of gate vibration was vortex formed by the failure of free shear layer at the bottom edge of gate. Based on the mechanism of flow-induced vibration, Naudascher [7] divided the gate vibration into three categories: (1) gate vibration induced by the instability of water flow; (2) gate vibration induced by flow instability and structure feedback mechanism; and (3) gate vibration caused by the induced force generated by structure motion. Thang [8,9] believed that gate vibration was caused by the downstream vortex resonance. Ishii et al. [10,11] supposed that fluid feedback force induced by gate vibration mainly included two aspects: pressure pulsation caused by the alternating flow and alternating excitation generated by the vortex at the gate bottom edge. Kunihiro et al. [12] considered that the vibration of radial gate with small opening was nonlinear self-excited vibration. In addition, Mostafiz et al. [13] and Keiko et al. [14] conducted studies on gate vibration from the perspective of frictional damping and gate opening-closing process in the physical model test. Kolkman et al. [15] expounded the fluid inertia mode of the gate vertical vibration and believed that, in self-excited vibration process, variation of the effective flow area led to the flow pulsation and fluid inertia may aggravate the gate vibration.

When gate vibrates by the hydrodynamic loads under the high-speed flow, flow field around the gate alters acutely and the structure dynamic loads change in turn, which forms the fluid–solid coupling effect. Most vibration control methods are to change the shape of the gate bottom edge, which could realize the stability of the flow pattern and weaken the vibration magnitude. For example, Markovic et al. [16] came up with a measure to effectively reduce hydrodynamic loads through placing penetrating orifice at the gate bottom beam. Erdbrink et al. [17] arranged some ventilation slots at the plane gate bottom and water could flow into the upstream panel and out at the gate bottom edge, which destroyed the reattachment effect between flow and gate bottom edge, smoothed the outflow state and effectively mitigate the vertical vibration. Seung et al. [18] proposed a method that can effectively reduce the gate vibration magnitude and abate the gate fatigue damage by setting a guide plate at gate bottom edge. Demirel et al. [19] mounted the horizontal porous baffle at elevation below the free surface to reduce the vortex magnitude in the downstream and enhance the stability of plane gate.

Due to the increasing demand for hydroelectric energy, construction of high-head water conservancy projects put forward a high requirement for the safe operation of gate. Much work so far has focused on the static load monitoring, vibration displacement, structural stress and so on. These achievements were largely obtained by fixed gate opening, and the flow under the gate can be viewed as a steady flow [20 23]. However, little attention has been devoted to the gate vibration and its effect on the holding force in closing process. When gate is in continuously closing process, fluid boundary deforms in a large scale and the flow passing the gate becomes markedly unsteady process. To determine the characteristics of gate vibration and holding force in closing process, we simultaneously measured the gate acceleration and holding force through the physical model test and then examined the correlation between the two variables. Further studies on the characteristics of vortex shedding in closing process and the method for evaluating the gate vibration effect on the holding force will be summarized in our subsequent study.

This paper is divided into four parts. Section 1 briefly describes the achievements about gate vibration and its control methods. Section 2 introduces the arrangement of physical model test and measuring point installation. Section 3 mainly discusses the characteristics of gate vibration and holding force and examines the correlation between gate vibration and holding force under two operation modes. The conclusion is given in Section 4.

2. Physical Model Test

The physical model test made of transparent Perspex plastic, at the scale 1:25, was established based on the gravity similarity criterion and flow similarity criterion. The parameters of the steel plane gate in the actual project are shown in Table 1. The physical model test was composed of five parts: tank, chamber, narrow-section, pressure hole and drainage channel (Figure 1). Two plane gates with same size were made of Perspex plastic and hydro-elastic material, respectively. Meanwhile, the hydro-elastic gate model should satisfy the structural similarity criterion to ensure the normal similitude of gate stiffness [24]. The mass and elastic modulus of hydro-elastic gate is 7.36 kg and 8 GPa, respectively.

Table 1. Parameters of the prototype gate.

Projects	Prototype Plane Gate
Spillway tunnel	6×14 m^2 (width × height)
Gate slot	8.4×1.6 m^2 (width × thickness)
Steel plane gate	$7 \times 14.2 \times 1.4$ m^3 (width × height × thickness)
Mass	100 t
Operation speed	1.5 m/min
Support type	Sliding block
Seal	Downstream seal
Gate Bottom edge	Composite type (upstream inclination angle 50° and down inclination angle 30°)
Operation requirement	Closing in transient flow and opening in still water
Elastic modulus E	200 GPa
Poisson's ratio μ	0.3
density ρ	7.85×10^3 kg/m^3

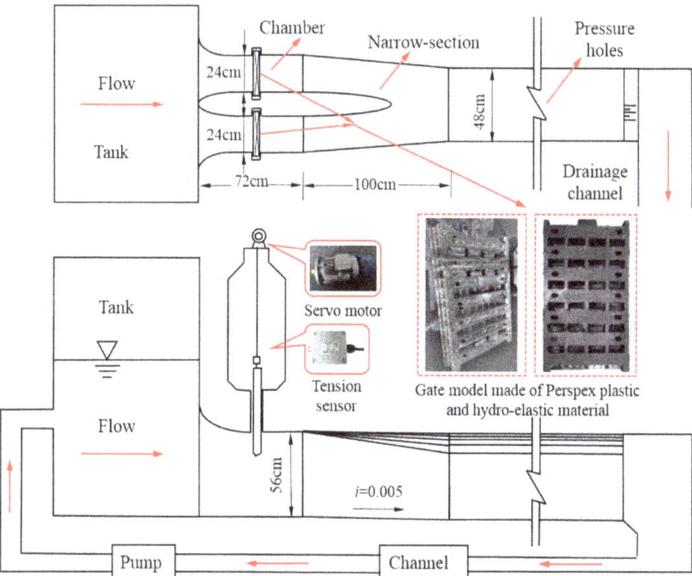

Figure 1. Schematic diagram of experiment setup.

In the physical model test, the plane gate was connected to the servo motor by steel cable and gate operation speed was 0.005 m/s. The tension sensor with a range of 100 kg was placed to measure the holding force. Three acceleration sensors were installed near the gate lug with lateral direction (gate width direction), vertical direction (gate height direction) and horizontal direction (flow direction), successively. Pulsating pressure sensors were arranged on the gate panel and bottom edge. Meanwhile, in the case of fixed gate opening the recording of data lasted for 60 s. In the case of closing process data acquisition lasted for the duration of the operation. Data Acquisition & Signal Processing (DASP) intelligent acquisition system developed by China Orient Institute of Noise & Vibration was employed to acquire and process the data cooperated with constant current power supplier. It can simultaneously measure the gate vibration acceleration and holding force at the sampling frequency of 200 Hz [25]. Data measuring, transmission and processing are depicted in Figure 2. Besides, in the physical model test, there are two operation modes of the gate, namely fixed gate opening and closing process. Gate opening ratio e in fixed gate opening equals gate opening height dividing the height of spillway tunnel. The upstream water depths were 0.64, 0.72, 0.8 and 0.92 m in the physical model test, successively (prototype depths were 16, 18, 20 and 23 m, respectively). Downstream outflow condition is free outflow. Rectangular weir was arranged in front of the drainage channel to measure the flow rate. The water levels on the tank and the rectangular weir were measured through water level meters. In addition, in closing process, we obtained the relationship acceleration or holding force with time from the experiment. To get the vibration or holding force at a certain gate opening in closing process, we firstly assume that gate opening is linear with the time. Subsequently, the vibration or holding force is approximately equal to the results of a specific time interval in closing process. In this paper, the specific time is linear with a certain gate opening and the interval is set to 1 s with 200 data.

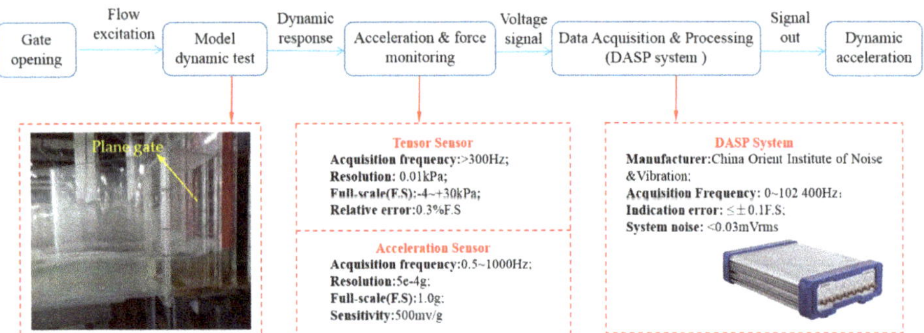

Figure 2. Dynamic testing system applied in the physical model test.

3. Results and Discussion

3.1. Characteristics of Vibration

Characteristics of plane gate vibration are related to the constraint boundaries and external excitations. Constraint boundaries of plane gate mainly include steel cable connected to the hoists, friction contact between the slide block and track, and friction contact between the rubber seal and sidewall. Once the constraint condition is determined, gate vibration is related to the external excitations. Thus, it is necessary to clarify the relationship between gate vibration with gate opening and upstream water depth. In previous investigations, plane gate was regarded as a rigid body, and its vertical vibration can be equivalent to a single degree of freedom [26,27]. Elastic deformation of steel cable is the main manifestation of the structural stiffness. Given the generalized stiffness theory, the equivalent stiffness of steel cable equals the generalized force dividing generalized displacement. When steel cable has the axial force P, axial deformation Δ can be expressed

$$\Delta = \frac{PL}{EA} \tag{1}$$

The equivalent stiffness of the steel cable K is, that is the gate vertical equivalent stiffness,

$$K = \frac{P}{\Delta} = P/(\frac{PL}{EA}) = \frac{EA}{L} \tag{2}$$

Thus, nature frequency of simplified system ω_n can be expressed

$$\omega_n = \sqrt{\frac{K}{m}} = \sqrt{\frac{EA}{Lm}} \tag{3}$$

where E is the elastic modulus of steel cable; A is the sectional area of the steel cable; L is the steel cable length; and m is simplified system mass.

On basis of the equivalent stiffness theory, gate vertical equivalent stiffness K is related to the steel cable length L and simplified system mass m. Through the physical model test, we mainly analyzed and compared the characteristics of gate vibration under the two operation modes.

3.1.1. Vibration in Fixed Gate Opening

Because the acceleration spectrums are similar under different water depth conditions, Figure 3 only depicts the acceleration spectrums under three directions for the upstream water depth of 20 m. As revealed in Figure 3, the acceleration in three directions decreases gradually with the gate opening decrease, which is consistent with the prototype observation by Yan et al. [28]. Yan et al. pointed out that flow pulsation pressure decreased with gate opening decrease and root mean squire (RMS) of pulsation pressure can reach 10% of the upstream total head in large gate opening. However, Yang et al. [29], with Particle Image Velocimetry (PIV) technology through the physical model test, stated that, the smaller is the gate opening and the greater is the flow rate, the greater is the gate vibration. It was confused that gate vibration unexpectedly emerged different results at the same gate opening. The gate operation condition might account for the difference. In this physical model test and prototype observation from [28], the flow passing the gate is free outflow and not restricted by the downstream water depth. Nevertheless, gate vibration has been influenced by the downstream water depth in the physical model test in [29]. This means that different outflow conditions could produce different results. Thus, outflow condition is also an effect factor of gate vibration.

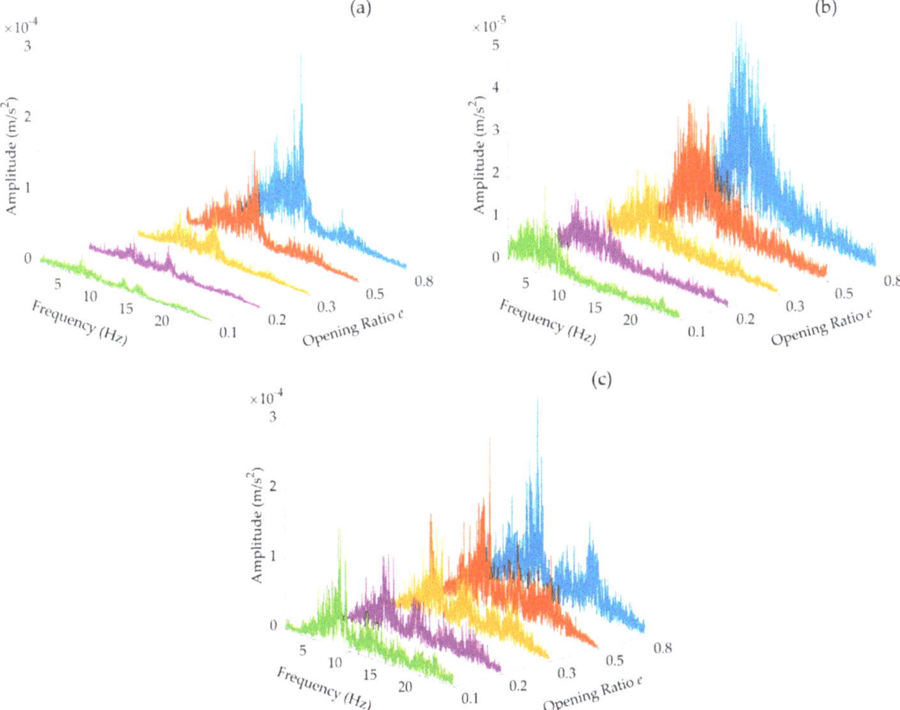

Figure 3. Acceleration spectrums in three directions for upstream water depth of 20 m: (**a**) lateral direction; (**b**) vertical direction; and (**c**) horizontal direction.

Furthermore, the dominant frequency of vibration in three directions was also obtained from the results in Figure 3. In lateral direction, the dominant frequency is mainly concentrated around 8 Hz. As gate opening decreases, the frequency is not prominent and vibration energy is dispersed gradually. Then, in vertical direction, the dominant frequency gradually decreases with gate opening decrease and is distributed up to 10 Hz. Vibration energy is relatively concentrated with a peak value and there is no energy dispersion. In horizontal direction the dominant frequency with multiple peaks under each gate opening has no obvious change. The frequency is mainly distributed below 20 Hz and frequency bandwidth is relatively wide. Besides, comparing the vibration spectrums at different water depths, we found upstream water depth could markedly change the acceleration magnitude and have a little influence on the dominant frequency yet. Meanwhile, flow pulsation frequency near the gate bottom edge is relatively small. The dominant frequency is 0.13–0.2 Hz. Because vibration frequency is far away from the flow pulsation frequency, the plane gate does not resonate with the flow and is self-excited vibration in fixed gate opening.

3.1.2. Vibration in Closing Process

In closing process, this paper pays more attention to the gate vertical vibration. Figure 4 provides the vertical acceleration in closing process and its partial magnification diagram. Considering the characteristics of the vertical vibration, the gate vibration process can be divided into three stages: initial stage, transitional stage and crawl stage. In the initial stage, vibration acceleration is relatively small, and its regularity is insignificant. In the transitional stage, vibration acceleration increases gradually and gate continues to fall without intermittent motion. Subsequently, in the crawl stage, vibration acceleration increases continuously and gate drops down intermittently.

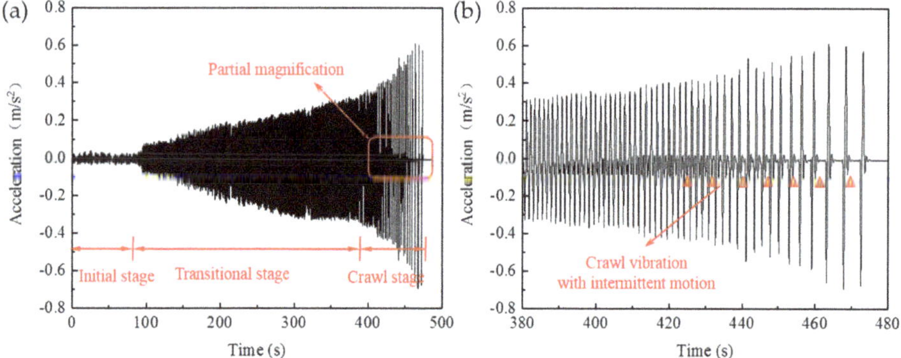

Figure 4. Vertical acceleration with gate-closing time: (**a**) vertical acceleration in closing process; and (**b**) partial magnification of vertical acceleration.

Figure 5a shows the amplitude of vertical acceleration in closing process. The amplitude is different for different stages. In the initial stage, the amplitude varies slightly and maintains at 0.01 m/s². Afterwards, the amplitude increases gradually. The higher the upstream water depth, the greater the amplitude. The maximum amplitude can reach 0.11 m/s² approximately in the crawl stage and is almost 10 times higher than the initial stage. Moreover, Figure 5b plots the dominant frequency of the vertical vibration in closing process. For different upstream water depths, frequency distribution is relatively stable, and its regularity is remarkable after the initial stage. In this period, the dominant frequency gradually decreases from 14 to 7 Hz in the form of steps. As upstream water depth increases, the step moves into the origin of the coordinate and the frequency decreases gradually. This means gate vibration is directly related to the upstream water depth. Due to the dominant frequency far away from the flow pulsation frequency, we regard that gate vertical vibration is self-excited vibration in closing process.

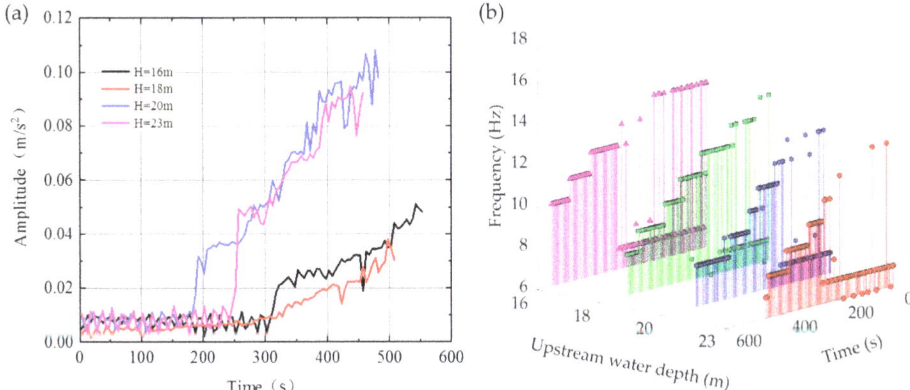

Figure 5. Characteristics of the vertical vibration in closing process: (**a**) amplitude of acceleration; and (**b**) dominant frequency of acceleration.

In addition, Figure 6 reveals the amplitude of acceleration in lateral and horizontal direction. The characteristics of lateral vibration is the same as that of vertical vibration. Nevertheless, the vibration magnitude is relatively small, and the maximum of amplitude can reach 0.013 m/s^2. In horizontal direction, the amplitude changes a little with time and mostly maintains below 0.008 m/s^2. However, unusual fluctuation peaks of acceleration appeared near 130 s, as depicted in Figure 6a. The main reason for the situation lies in the defect of the model itself. Because the physical model was built in summer and fixed by steel structure, the guide track deformed slightly at a lower temperature. The defective area is located in large gate opening. Gate opening ratio e is roughly 0.83 and it is not the focus area of the study. Moreover, this defect does not significantly influence the vertical vibration results for a small gate opening, as described in Figure 5a. Therefore, we can ignore this defect in the following research.

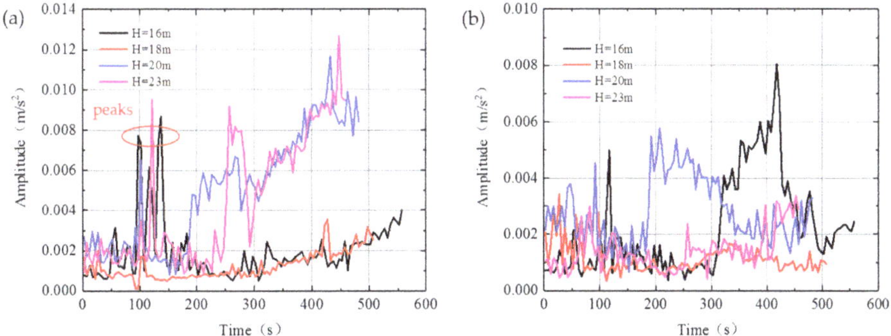

Figure 6. Characteristics of acceleration amplitude in closing process: (**a**) lateral direction; and (**b**) horizontal direction.

3.1.3. Comparison of Vibration in Two Operation Modes

The above describes the characteristics of gate vibration in fixed gate opening and closing process, respectively. This section mainly compares gate vibration for two operation modes. When gate is in fixed gate opening, the acceleration in the three directions decreases gradually with gate opening decrease. The lateral and horizontal vibrations are larger than the vertical vibration. However, in closing process, acceleration increases gradually with gate opening decrease, which is contrary to the fixed gate opening. Besides, vertical acceleration is the largest, followed by the lateral direction and horizontal direction is the smallest. From the perspective of the frequency, vibration frequency is mainly related to the gate opening and is independent of the upstream water depth in fixed gate opening. Frequency in each direction is markedly different. Nevertheless, in closing process, upstream water depth and gate opening both impact the vibration frequency. As the upstream water depth increases and gate opening decreases, vertical vibration frequency reduces gradually. The above differences are mainly caused by two aspects: the gate motion state and unsteady flow. Firstly, the gate motion state influences the boundary conditions and nature frequency of simplified system. Secondly, unsteady flow arising from gate-closing process significantly changes the hydrodynamic loads acting on the gate body. These two effects motivate a strong fluid–solid coupling in closing process, and it is markedly different from that of fixed gate opening.

3.2. Characteristics of Holding Force

Gate holding force directly determines the hoists capacity and is of great significance to the stable and safe operation of gate. Many accidents are related to the holding force in practical engineering. To determine the hoists capacity, hydrodynamics loads is a vital influential factor, in addition to the mechanical factors. Although the influence of hydrodynamics loads on hoists capacity has been explained in the relevant criterion [30], the criterion does not seem to meet the practical requirements on condition of the complex gate structure, operation conditions and design scale at present. It is necessary to clarify the hydrodynamics loads in closing process. The holding force F_{hoist} can be expressed:

$$F_{hoist} = nG_{gate} + G_j + W_s + P_X - P_t - F_f \tag{4}$$

where n is modified coefficient of gate weight, G_{gate} is gate self-weight, G_j is additional weight, W_s is water column weight on the gate top, F_f is friction force and $P_X - P_t$ is vertical hydrodynamic force at the gate bottom edge.

3.2.1. Holding Force in Two Operation Modes

From Equation (4), many factors can influence the holding force. In fixed gate opening, water column on the gate top and panel thrust have a relatively clear theoretical relationship with gate opening and upstream water depth. Thus, the characteristics of holding force mainly represent the characteristics of the pressure at gate bottom edge.

Table 2 summarizes the holding force and its standard deviation in fixed gate opening. As upstream water depth increases, holding force generally decreases. In large gate opening, however, the holding force increases slightly with the upstream water depth increase because gate lattice is partly filled with water. The higher is the water level, the more water there is in the gate lattice and the greater is the holding force. In small gate opening, gate lattice is fully filled with water for different operation conditions. The influence of water in the gate lattice is insignificant for the holding force.

Table 2. Holding force and its standard deviation in fixed gate opening.

Opening Ratio e	Holding Force (kN)				Standard Deviation			
	16 m	18 m	20 m	23 m	16 m	18 m	20 m	23 m
0.8	2391.968	2437.359	2556.328	2513.437	0.182	0.216	0.289	0.346
0.5	2324.772	2097.109	2008.719	1828.062	0.180	0.188	0.194	0.154
0.3	2027.556	1781.109	1778.297	1114.462	0.176	0.174	0.139	0.132
0.2	1864.470	1469.297	1517.547	678.750	0.162	0.134	0.137	0.126
0.1	1371.220	721.812	104.765	477.906	0.133	0.130	0.132	0.124

In addition, standard deviation gradually decreases with the gate opening decrease for a certain upstream water depth, as shown in Table 2. Standard deviation is proportional to water depth in large gate opening. In small gate opening, it is inversely proportional to water depth. It indicates that upstream water depth impacts the pressure at gate bottom edge and the influence is nonlinear. Therefore, the suggestion that the pressure specified as 20 kN/m or a linear formula in the criterion does not seem to satisfy the practical requirements and need to further investigate [30].

For different upstream water depths, dominant frequency of holding force in each opening is 0.033–0.217 Hz. The frequency is not equal or close to the vertical vibration frequency and is not close to the flow pulsation frequency yet.

Subsequently, Figure 7 plots the holding force for different conditions in closing process. As upstream water depth rises, the gate-closing time decreases gradually when gate stops moving. In the initial stage, holding force varies slightly and the maximum holding force occurs near 100 s. It indicates that vertical hydrodynamic loads varies significantly and can markedly change the holding force. In the transitional crawl stage, holding force decreases gradually in a reciprocation form. It is consistent with the existing research [31]. Liang et al. [31] reckoned that holding force is a process of elastic energy stored by the steel cable releasing and storing. Specifically, the elastic energy of steel cable goes through a cycle of release–reserve–re-release–re-reserve in closing process. Afterwards, the friction forces by the loads acting on the gate body gradually increase, make the holding force gradually decrease and prevent the gate from dropping down.

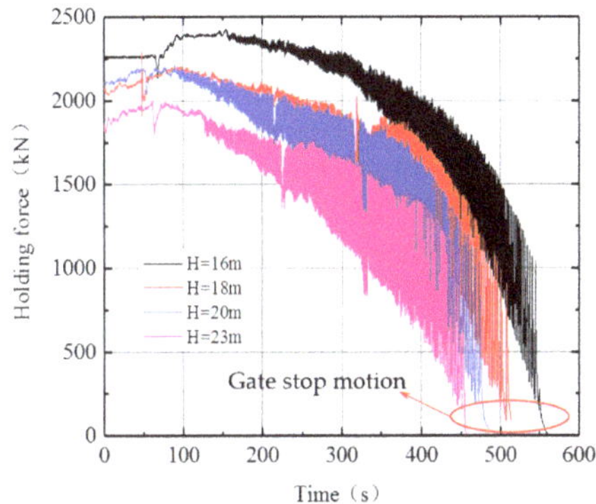

Figure 7. Holding force for different conditions in closing process.

Figure 8a shows the fluctuation value of holding force in closing process. In the initial stage, the fluctuation value is small. As gate continues to go down, the fluctuation value suddenly increases rapidly. The higher is the upstream water level, the greater is the fluctuation value. Besides, Figure 8b depicts the dominant frequency of holding force. Frequency distribution is regular and its value reduces from 14 to 7 Hz in a step form. In addition, without considering the elastic hysteresis of the steel cable, axial deformation of steel cable obtained through the Equation (1) should equal the gate vertical vibration displacement. Meanwhile, we noticed that the dominant frequency of holding force is consistent with the gate vertical vibration frequency in closing process, which indicates that it is reasonable to analyze the vertical vibration through the holding force. It can be further manifested that it is feasible, in a study of gate vertical vibration, to simplify gate vertical vibration to a single degree of freedom vibration system.

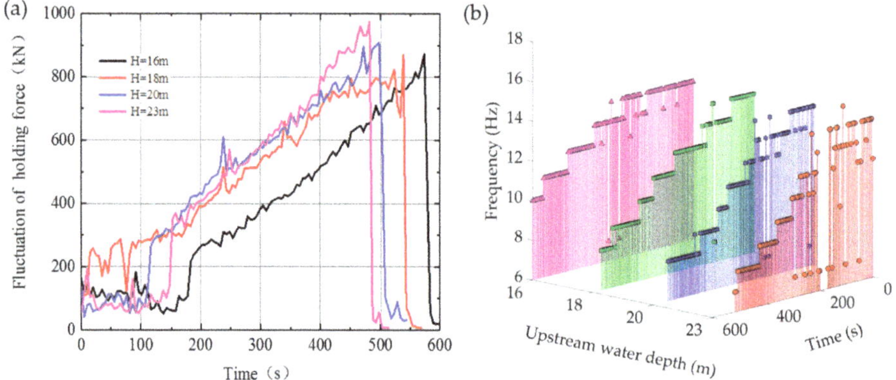

Figure 8. Characteristics of holding force in closing process: (**a**) fluctuation value of holding force; and (**b**) dominant frequency of holding force.

3.2.2. Comparison of Holding Force in Two Operation Modes

Figure 9 plots the comparison results of the average of holding force for two operation modes. Overall, holding force in fixed gate opening is greater than the closing process under the same gate opening. With increase of the upstream water depth, difference of holding force between the two operation modes is large in large gate opening and is small in small gate opening. The reason for the situation is that in closing process the cross-section of flow decreased in initial stage, which led to the upstream water level to rise gradually. The water level rising caused the horizontal panel thrust, and the friction force between gate and the track increased, which made holding force less than the fixed gate opening. The higher is the upstream water level, the greater is the difference. However, when the gate dropped down intermittently in the crawl stage, the water level rising was no longer apparent, and the difference was insignificant at small gate opening.

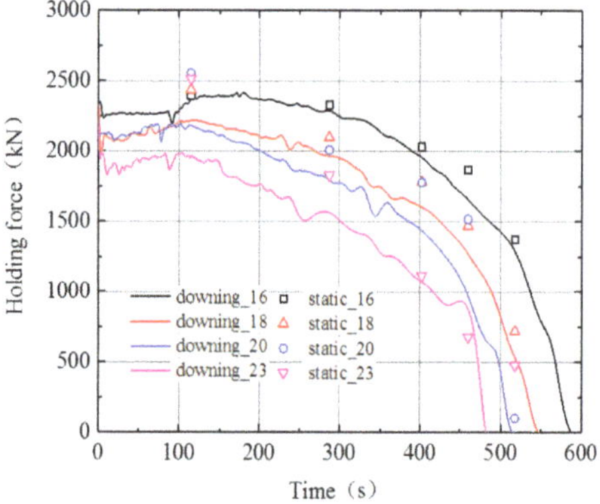

Figure 9. Average of holding force under the two operation modes.

Next, the dominant frequency of holding force was analyzed for two operation modes. In fixed gate opening, the dominant frequency of holding force is 0.2 Hz and energy spectral frequency is mainly concentrated below 1 Hz. Nevertheless, vertical vibration frequency is mainly concentrated around 5–8 Hz. The frequency of holding force is inconsistent with the vertical vibration frequency and is close to the flow pulsation frequency yet. In closing process, the vertical vibration frequency and the frequency of holding force are consistent. The frequency decreases from 14 to 7 Hz gradually.

Thus, the frequency between the holding force and vibration is not same for two operation modes. The difference may be attributed to the friction force caused by the gate motion state. In fixed gate opening, plane gate is in a state of static friction with the track. The direction of static friction force is changeable with the gate motion tendency. We tentatively put forward that, when the hydrodynamic loads are converted into the steel cable tension, static friction force with variable direction has a certain filtration effect on the loads. Only if the hydrodynamic loads are greater than the maximum static friction force can the loads be converted into the steel cable tension. Otherwise, the loads will be balanced through the static friction force. Thus, frequency of holding force is consistent with the flow pulsation frequency in fixed gate opening. Compared to the fixed gate opening, plane gate is in a state of sliding friction, and the friction force direction, opposite to the gate motion, is constant in closing process. Thus, the frequency of holding force is consistent with vertical vibration frequency in this case.

3.2.3. Crawl Vibration in Closing Process

In the closing process, the length of steel cable, water weight in gate lattice and water column on the gate top increase gradually. These changes directly reflect the increase of length L and mass m in Equation (3), which makes the nature frequency of the simplified system decrease. Besides, the flow boundary deformation and unsteady flow under the gate result in the variation of fluid inertia and make flow pulsation pressure acting on the gate bottom edge intensify significantly. Cai et al. [23] studied that the seismic waves had influenced on the maximum hydrodynamic pressure in curved gate. The results indicate that maximum hydrodynamic pressure of gate panel, considering the fluid–solid coupling effect, was 4–5 times higher than the theoretical value by the additional mass method. Thus, strong fluid–solid coupling effect could increase the hydrodynamic loads and crawl vibration appears in small gate opening. Ji et al. [32] regarded that the difference of dynamic and static friction coefficient was the main factor for crawl vibration. The necessary condition for the vibration is given as follows: $\Delta\mu \times P/(v \times \sqrt{mk}) > 1$ (where $\Delta\mu$ is the difference of dynamic and static friction coefficient, P is the horizontal panel thrust, v is the closing gate speed, m is the gate mass and k is the transmission rod stiffness).

Table 3 provides the time parameters about the crawl vibration in the physical model test. With the upstream water depth increase, the gate-closing time decreases and duration of crawl vibration increases gradually. We regard that crawl vibration should be related to the upstream water depth and happen at specific external loads. When the gate structure and upstream water depth are determined, the horizontal panel thrust can be viewed as a control index for the crawl vibration. From the results in [32], horizontal panel thrust also played an important role in the crawl vibration. Once panel thrust exceeds the critical threshold, crawl vibration will appear. As upstream water depth increases, the time for panel thrust reaching the threshold decreases gradually so that initial crawl time decreases, as shown in Table 3.

Table 3. The time parameters about the gate crawl vibration. Unit: s.

Upstream Water	16 m	18 m	20 m	23 m
Gate closed time	115.3	107.6	100.3	95.78
Initial crawl time	95.05	87.13	73.78	69.81
Duration of crawl vibration	20.25	20.47	26.52	25.96

After the gate intermittent motion, the gate stopped moving at a certain distance from the chamber floor. The higher is the upstream water depth, the larger is the distance. The gate is not fully closed and the failure of gate-closing appears, which seriously threatens the operation of the projects. The failure has also happened in other physical model tests and engineering practices [33,34]. Because steel cable cannot provide the vertical downward loads, the gate can only utilize the self-weight, additional weight, water column on the gate top and water weight in the gate girder to close. Especially for the heavily sediment-carrying river, the sediment deposited in the gate slots exacerbates the above phenomenon [33]. Liang et al. [31] and Novak et al. [35] both suggested increasing gate speed can improve gate stability, avoid or delay the crawl vibration to a certain extent, which was good for the gate-closing.

To solve the above problem, increasing vertical downward loads by modified gate shape or additional mass is a common technical method. Ma et al. [34] proposed the method shown in Figure 10a. Placing a protruding boundary in front of the gate bottom edge and making full use of the water column could increase the vertical downward loads. Figure 10b provides the holding force for three body shapes in the physical model test. After the body modification, the holding force gradually increases, and gate can be completely closed. Besides, we should also focus on the concerned problems caused by body modification. In Figure 10b, when gate opening ratio e is 1–0.5 (gate-closing time is less than 250 s), the fluctuation value of holding force for three body shapes is relatively small. When gate opening ratio e is less than 0.5, the fluctuation value rapidly increases. The fluctuation value for the gate of leading edge with convex is roughly three times as much as the original body. The reason for this situation is that the gate of leading edge protrusion increases effective bottom width and leads to the separation and re-attachment of flow through the gate bottom intensified. It indicates the gate bottom width is direct factor for pressure fluctuation. Moreover, because hoists capacity converted by the prototype is about 500 N in the physical model test, the holding force has exceeded the maximum capacity for the modified gate of leading edge with convex. Fluctuation value of holding force goes back and forth to a large extent, which performs the axial deformation of steel cable and stress concentration in gate lug area. It is worth noting that stress concentration may result in tearing failure of lifting lug. Thus, the characteristics of holding force should be given more attention to prevent adverse problems in the actual operation of the modified gate.

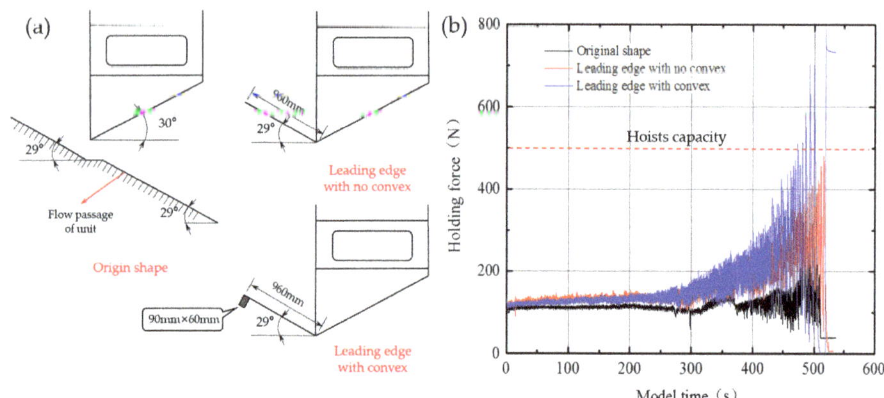

Figure 10. Measures to make gate be closed completely in the physical model test: (**a**) original and modified shape of gate bottom edge [34]; and (**b**) holding force for three body shapes.

3.3. Correlation between Gate Vibration and Holding Force

Through the physical model test, we simultaneously measured the gate vibration and holding force. Gate vibrations in three directions are independent of each other. To explore the correlation between the gate vibration and holding force, we directly conducted the correlation and significance test between the two variables under two operation modes, respectively.

3.3.1. Correlation in Fixed Gate Opening

Table 4 shows the correlation and significance test between gate vibration and holding force in fixed gate opening. The distribution rule of correlation coefficient is not evident in three directions. Moreover, absolute value of correlation coefficient is relatively small and less than 0.3. It suggests that the correlation between gate vibration and holding force is weak. Thus, we can believe that gate vibration has no obvious impact on the holding force and its influence can be neglected for fixed gate opening.

Table 4. Correlation between gate vibration and holding force in fixed gate opening.

Direction	Water Depth (m)	Gate Opening Ratio e				
		0.1	0.2	0.3	0.5	0.8
Lateral	16	−0.257 **	0.015	−0.225 **	0.053 **	−0.043 **
	18	−0.127 **	0.169 **	0.142 **	0.016	−0.053 **
	20	−0.134 **	−0.156 **	−0.256 **	0.185 **	−0.173 **
	23	−0.249 **	−0.177	−0.088 **	0.173 **	0.458 **
Vertical	16	−0.472 **	0.057 **	0.049 **	0.085 **	0.108 **
	18	−0.031 **	0.228 **	0.187 **	−0.267 **	0.053 **
	20	−0.189 **	−0.109 **	−0.546 **	−0.095 **	−0.082 **
	23	−0.186 **	−0.201 **	0.093	−0.166 **	−0.125 **
Horizontal	16	0.169 **	−0.005	−0.009	0.002	−0.041 **
	18	−0.027	−0.054 **	0.248 **	−0.066 **	−0.014
	20	−0.03 **	0.001	−0.123 **	−0.179 **	−0.034 **
	23	−0.018	−0.022 *	0.014	−0.022 *	−0.225 **

Note: * denotes a significant difference ($p < 0.05$), ** denotes a significant difference ($p < 0.01$).

3.3.2. Correlation in Closing Process

Table 5 describes the correlation and significance test between gate vibration and holding force in closing process. The results of significance test indicate a significant correlation between gate vibration and holding force. The correlation coefficient is different for different stages. To be specific, in the initial stage correlation coefficient in three directions is relatively small and the correlation is weak, indicating that gate vibration has a little influence on the holding force and can be ignored for calculating the hoists capacity. In the transitional stage, the average of correlation coefficient in lateral, vertical and horizontal direction is −0.565, 0.944 and −0.764, respectively. According to the correlation coefficient from large to small, they were ordered as: vertical > horizontal > lateral. Subsequently, in the crawl stage, the average of correlation coefficient in lateral, vertical and horizontal direction is −0.723, 0.733 and −0.664, respectively. The correlation coefficient from large to small was as follows: vertical > lateral > horizontal. Furthermore, in the transitional and crawl stage, holding force is positively correlated with the vertical vibration and is negatively correlated with the lateral and horizontal vibrations yet. It seems that gate motion state can account for the difference. In the transitional stage, the gate keeps falling and the vertical vibration is directly represented in the holding force. On the basis of the Newton's second law, there is a liner relationship between gate vibration and holding force. Thus, the correlation is significantly positive, and its coefficient can reach 0.944. In the crawl stage, because the gate is in a state of intermittent motion, correlation coefficient is 0.733 and smaller than the transitional stage. Meanwhile, lateral and horizontal vibrations make the friction force originating from the seal pressure and panel thrust increase. Thus, the correlation coefficient is negative, and the vibration has a mitigating effect on the holding force. In contrast, if the gate is in an opening process, lateral and horizontal vibrations can increase the holding force and the correlation coefficient should be positive. In summary, the gate vibration effect on the holding force in closing process should be considered in determining the hoists capacity.

In addition, we concentrate only on the integral gate vibration as a rigid body in closing process. The limitation of this study is not to consider the horizontal deformation of gate itself in closing process. Anami et al. [36] replaced the circular-arc plate with a vertical, flat, rigid weir-plate and analyzed the rotational vibration along the weir-plate surface through the potential flow theory. The simplification of curved gate manifested that gate vibration was different at different gate heights in fixed gate opening. Yang et al. [29] also found vibration displacement at the bottom of plane gate was greater than at the middle in fixed gate opening. To facilitate the layout and not damage the structure of hydro-elastic gate, however, three acceleration sensors were installed near the gate lug in this study. Thus, the vibration near the lug might be unable to fully represent the characteristics of horizontal vibration. Nevertheless, the structure and operation mode between the curved gate and plane gate are both different. The reliability of the above results applied to a moving plane gate in this paper is uncertain. We have thus far not been able to quantity the uncertainty through the existing results. The following research will focus on the characteristics of horizontal vibration at different gate heights in closing process.

Table 5. Correlation between gate vibration and holding force in closing process.

Water Depth (m)	Initial Stage			Transitional Stage			Crawl Stage		
	Lateral	Vertical	Horizontal	Lateral	Vertical	Horizontal	Lateral	Vertical	Horizontal
16	0.045 **	0.418 **	0.021	−0.472 **	0.946 **	−0.644 **	−0.724 **	0.737 **	−0.644 **
18	−0.227 **	0.761 **	−0.374 **	−0.566 **	0.955 **	−0.874 **	−0.731 **	0.723 **	−0.650 **
20	0.271 **	0.285 **	−0.032 *	−0.555 **	0.941 **	−0.678 **	−0.720 **	0.732 **	−0.664 **
23	−0.138 **	0.389 **	−0.041 **	−0.667 **	0.936 **	−0.862 **	−0.720 **	0.741 **	−0.698 **

Note: * denotes a significant difference ($p < 0.05$), ** denotes a significant difference ($p < 0.01$).

4. Conclusions

Physical model test was used to investigate the characteristics of gate vibration and holding force in a plane gate at two operation modes. More attention was paid to the gate vertical vibration and the correlation between the gate vibration and holding force. The results are as follows.

The vibration acceleration is significant in lateral and horizontal directions, and it is 10 times higher in the vertical direction in fixed gate opening. In closing process, however, vertical vibration is stronger than those in the lateral and horizontal directions. Dominant frequency of vertical vibration gradually decreases from 14 to 7 Hz in a step form. The experiment results also show that gate vertical vibration simplified into single degree of freedom is reasonable through equivalent stiffness theory.

At the same gate opening, the holding force in closing process is less than the fixed gate opening. The horizontal panel thrust of plane gate can be treated as a hydraulic factor for the crawl vibration. Once the panel thrust exceeds the critical threshold, crawl vibration happens. After the crawl vibration, the gate stops motion at a certain distance from the chamber floor and the failure of gate-closing happens. The higher is the upstream water level, the greater is the distance. Increasing the vertical downward loads by modified shape of gate bottom edge can solve the failure of gate closing. More attention should be paid to the holding force to prevent the large fluctuation value from exceeding hoists capacity and tearing failure.

For fixed gate opening, gate vibration has almost no influence on holding force and can be ignored in determining the hoists capacity. In closing process, however, absolute value of correlation coefficient in three directions is relatively high. Holding force is significantly positively correlated with the vertical vibration and is significantly negatively correlated with lateral and horizontal vibrations. Thus, the gate vibration effect on the holding force should be considered in determining the hoists capacity under closing process.

Author Contributions: Conceptualization, Y.W. and G.X.; methodology, Y.W. and G.X.; investigation, Y.W.; data curation, W.L.; writing—original draft preparation, W.L.; writing—review and editing, F.L. and Y.D.; visualization, Y.W.; and supervision, G.X. All authors have read and agreed to the published version of the manuscript.

Funding: This research was funded by National Natural Science Foundation of China (Grant No. 51779166).

Acknowledgments: The authors gratefully acknowledge the editors and anonymous reviewers for their insightful and professional comments, which greatly improved this manuscript.

Conflicts of Interest: The authors declare no conflict of interest.

References

1. Darbre, G.R.; Smet, C.A.M.D.; Kraemer, C. Natural frequencies measured from ambient vibration response of the arch dam of Mauvoisin. *Earthq. Eng. Struct. D.* **2015**, *29*, 577–586. [CrossRef]
2. Wójcicki, Z.; Kostecki, S.; Grosel, J. Operational Modal Analysis of Weir on Odra River in Poland. *Procedia Eng.* **2016**, *153*, 874–881. [CrossRef]
3. Peng, Y.; Zhang, J.; Xu, W.-L.; Rubinato, M. Experimental Optimization of Gate-Opening Modes to Minimize Near-Field Vibrations in Hydropower Stations. *Water* **2018**, *10*, 1435. [CrossRef]
4. Lian, J.; Zheng, Y.; Liang, C.; Ma, B. Analysis for the Vibration Mechanism of the Spillway Guide Wall Considering the Associated-Forced Coupled Vibration. *Appl. Sci.* **2019**, *9*, 2572. [CrossRef]
5. Hardwick, J.D. Flow-induced vibration of vertical-lift gate. *J. Hydraul. Div. ASCE.* **1974**, *1005*, 631–644.
6. Jongeling, T.H.G. Flow-induced self-excited in-flow vibrations of gate plates. *J. Fluids Struct.* **1988**, *2*, 541–566. [CrossRef]
7. Naudascher, E. Flow-induced loading and vibration of gates. *Proc. Int. Symp. Hydraul. High Dams* **1988**, 1–18.
8. Thang, N.D.; Naudascher, E. Self-excited vibrations of vertical-lift gate. *J. Hydraul. Res.* **1986**, *24*, 391–404. [CrossRef]
9. Thang, N.D. Gate Vibrations due to Unstable Flow Separation. *J. Hydraul. Eng.* **1990**, *116*, 342–361. [CrossRef]
10. Ishii, N.; Knisely, C.W. Flow-induced vibration of shell-type long-span gates. *J. Fluids Struct.* **1992**, *6*, 681–703. [CrossRef]
11. Ishii, N.; Knisely, C.W.; Nakata, A. Coupled-Mode Vibration of Gates with Simultaneous Over-and Underflow. *J. Fluids Struct.* **1994**, *8*, 455–469. [CrossRef]
12. Ogihara, K.; Ueda, S. The Conditions of Self-Excited Vibration Occurring in Radial Gates. *Water Power Conf.* **1999**, 1–10. [CrossRef]
13. Chowdhury, M.R.; Hall, R.L. Dynamic Performance Evaluation of Gate Vibration. *J. Struct. Eng.* **1999**, *125*, 445–452. [CrossRef]
14. Anami, K.; Ishii, N.; Knisely, C.W.; Tsuji, T.; Oku, T.; Sato, S. Friction-Maintained Dynamic Stability. In Vibration Problems ICOVP 2011. *Springer Proc. Phys.* **2011**, 779–785. [CrossRef]
15. Kolkman, P.A.; Vrijer, A. Vertical Gate Vibrations by Galloping or By Fluid Inertia. *J. Hydraul. Res.* **1987**, *25*, 418–423. [CrossRef]
16. Markovic, J.; Drobir, H. A high head gate innovation numerical and experiment analysis of hydrodynamic forces. *Math. Model. Civ. Eng.* **2010**, *2*, 27–34.
17. Erdbrink, C.D.; Krzhizhanovskaya, V.V.; Sloot, P. Reducing cross-flow vibrations of underflow gates: Experiments and numerical studies. *J. Fluids Struct.* **2014**, *50*, 25–48. [CrossRef]
18. Lee, S.O.; Seong, H.; Kang, J.W. Flow-induced vibration of a radial gate at various opening heights. *Eng. Appl. Comput. Fluid Mech.* **2018**, *12*, 567–583. [CrossRef]
19. Demirel, E.; Aral, M.M. A Design for Vortex Suppression Downstream of a Submerged Gate. *Water* **2020**, *12*, 750. [CrossRef]
20. Gao, Z.; Yan, G.; Liu, P.; Chen, F.; Lv, F. Study on Flow-Induced Vibration Characteristics of Hydraulic Hoist of large-Span up welling Radial Steel Gate. *Appl. Mech. Mater.* **2011**, *117*, 241–246.
21. Ng, F.C.; Abas, A.; Abustan, I.; Rozainy, Z.M.R.; Abdullah, M.; Jamaludin, A.b.; Kon, S.M. Fluid/structure interaction study on the variation of radial gate's gap height in dam. *IOP Conf. Ser. Mater. Sci. Eng.* **2018**, *370*, 012063. [CrossRef]
22. Shen, C.; Wang, W.; He, S.; Xu, Y. Numerical and Experimental Comparative Study on the Flow-Induced Vibration of a Plane Gate. *Water* **2018**, *10*, 1551. [CrossRef]
23. Cai, Y.F.; Han, Y.F. Research on the Radial Gate Fluid-Structure Coupling of Hydraulic Steel Gate. *Appl. Mech. Mater.* **2015**, *733*, 468–471. [CrossRef]
24. Lian, J.J.; Chen, L.; Liang, C.; Liu, F. Presentation and Verification of an Optimal Operating Scheme Aiming at Reducing the Ground Vibration Induced by High Dam Flood Discharge. *Int. J. Environ. Res. Public Health* **2020**, *17*, 377. [CrossRef] [PubMed]
25. Lian, J.J.; Chen, L.; Ma, B.; Liang, C. Analysis of the Cause and Mechanism of Hydraulic Gate Vibration during Flood Discharging from the Perspective of Structural Dynamics. *Appl. Sci.* **2020**, *10*, 629. [CrossRef]
26. Yang, M.; Lian, J.J.; Lin, J.Y. Excitation mechanism of flow induced plane gate vibration. *J. Hydrodyn.* **1997**, *4*, 437–449. (In Chinese)

27. Billeter, P.; Staubli, T. Flow-induced multiple-mode vibrations of gates with submerged discharge. *J. Fluids Struct.* **2000**, *14*, 323–338. [CrossRef]

28. Yan, G.H.; Chen, J.F. Prototype observation and study on flow induced vibration of curved gate in surface hole of spillway dam. *J. Hydroelectr. Eng.* **2012**, *2*, 142–147. (In Chinese)

29. Yang, T.; He, S.; Shen, C.; Yang, Q. Fluid-induced Vibrations Test of Hydraulic Plane Gate for Different Conditions. *Int. Conf. Sens.* **2016**, 1083–1089. [CrossRef]

30. Ministry of Water Resources of the People's Republic of China. *Specification for Design of Steel Gate in Hydraulic and Hydroelectric Engineering (NB 35055—2015)*; China Water & Power Press: Beijing, China, 2015; pp. 52–54. (In Chinese)

31. Liang, C.; Zhang, J.L.; Lian, J.J.; Liu, F. Research on slip-stick vibration of emergency gate induced by high dam flood discharge. *J. Hydraul. Eng.* **2018**, *49*, 1503–1511, 1522. (In Chinese)

32. Ji, S.K.; Zhou, L. Vibration analysis of sluice gate closure process for diversion of Yellow River. *Water Conserv. Hydropower Technol.* **2001**, *12*, 64–66. (In Chinese)

33. Li, G.Q.; Yang, J.W.; Yang, B.; Li, B. Analysis of prototype observation results of lifting and closing force of sluice gate in spillway tunnel of tianqiao hydropower station. *Water Conserv. Hydropower Technol.* **2005**, *10*, 4–7. (In Chinese)

34. Ma, C.; Sheng, C.; Lian, J.; Ma, B.; Liu, F. Failure analysis of a leaf gate jammed in closing process. *Eng. Fail. Anal.* **2020**, *110*, 104391. [CrossRef]

35. Novak, G.; Mlateralčnik, J.; Bombač, M.; Vošnjak, S. Hydrodynamic forces during the operation of a model radial gate. *J. Appl. Water Eng. Res.* **2016**, *51*, 70–77. [CrossRef]

36. Anami, K.; Ishii, N.; Knisely, C.W. Pressure induced by vertical planar and inclined curved weir-plates undergoing streamwise rotational vibration. *J. Fluids Struct.* **2012**, *29*, 35–49. [CrossRef]

© 2020 by the authors. Licensee MDPI, Basel, Switzerland. This article is an open access article distributed under the terms and conditions of the Creative Commons Attribution (CC BY) license (http://creativecommons.org/licenses/by/4.0/).

applied sciences

Article

Structural Health Monitoring of 2D Plane Structures

Behnam Mobaraki [1], Haiying Ma [2,*], Jose Antonio Lozano Galant [1] and Jose Turmo [3]

[1] Department of Civil and Building Engineering, Universidad de Castilla La Mancha,
13071 Ciudad Real, Spain; behnam.mobaraki@uclm.es (B.M.); joseantonio.lozano@uclm.es (J.A.L.G.)
[2] Department of Bridge Engineering, Tongji University, Shanghai 200092, China
[3] Department of Civil and Environmental Engineering, Universitat Politècnica de Catalunya BarcelonaTech,
08034 Barcelona, Spain; jose.turmo@upc.edu
* Correspondence: mahaiying@tongji.edu.cn

Abstract: This paper presents the application of the observability technique for the structural system identification of 2D models. Unlike previous applications of this method, unknown variables appear both in the numerator and the denominator of the stiffness matrix system, making the problem non-linear and impossible to solve. To fill this gap, new changes in variables are proposed to linearize the system of equations. In addition, to illustrate the application of the proposed procedure into the observability method, a detailed mathematical analysis is presented. Finally, to validate the applicability of the method, the mechanical properties of a state-of-the-art plate are numerically determined.

Keywords: observability method; structural system identification; plane strain analysis; 2D elements; structural health monitoring; inverse analysis; finite element method

Citation: Mobaraki, B.; Ma, H.; Lozano Galant, J.A.; Turmo, J. Structural Health Monitoring of 2D Plane Structures. *Appl. Sci.* **2021**, *11*, 2000. https://doi.org/10.3390/app11052000

Academic Editor: Hugo Filipe Pinheiro Rodrigues

Received: 18 January 2021
Accepted: 19 February 2021
Published: 24 February 2021

Publisher's Note: MDPI stays neutral with regard to jurisdictional claims in published maps and institutional affiliations.

Copyright: © 2021 by the authors. Licensee MDPI, Basel, Switzerland. This article is an open access article distributed under the terms and conditions of the Creative Commons Attribution (CC BY) license (https://creativecommons.org/licenses/by/4.0/).

1. Introduction

In recent years, the maintenance, health monitoring, and identification of structural systems are becoming more frequent all around the world [1,2]. A methodology to identify the mechanical properties of existing structures according to in situ measurements is defined as structural system identification [3] and is a necessary part of any Structural Health Monitoring system. In the maintenance and rehabilitation process of the structures, in addition to the in situ measurements and visual inspections, precise damage detection might be performed using numerical and analytical analysis [4,5]. Examples of non-model-based damage identification approaches can be found in [6,7].

Many structural system identification methods have been proposed for estimating the mechanical parameters of structures modelled with 1D elements, such as steel and concrete buildings as well as cable-stayed bridges, trusses, and frames [8–10]. However, despite the intricate nature of 2D structural models such as tunnels, dams, and culverts, few investigations have been conducted on the structural system identification of these structures [11–15]. For instance, geotechnical engineers who tried to estimate the behavior of buried structures encountered insufficient sets of input data [16,17]. In the case of segmental underground structures, both the constraining effect of the soil and the existence of the segment joints will generate problems for acquiring dynamic characteristics [18–20]. Khamsei et al. presented a new intelligent inverse analysis technique combining fuzzy systems, an imperialistic competitive algorithm, and numerical analysis for the back analysis of the Karaj Subway in Iran [21]. Dehghan et al. determined the geotechnical parameters using inverse analysis based on convergence data [22]. They also proved that their proposed method would be a more economical and time-saving technique in comparison to a design based on soil mechanic tests carried out by consultant engineers.

Structural system identification methods can be categorized as parametric and non-parametric [23–25]. With the quick increase in computer technology, jutting numerical software, and well-known packages such as MatLab or Mathematica, the popularity of non-parametric methods has increased. These methods define the transfer functions of a system, which implies that the input–output relation is characterized by a set of equations that may not have any explicit physical meaning. Vardakos et al. estimated the potential use of the differential evolution genetic algorithm in the back analysis of a tunnel response to obtain improved estimates of the model parameters by matching the model prediction with the monitored response [25].

Xiang et al. proposed two algorithms by means of automatically generating optimal measurements [26]. They proved the validity of their methods with academic examples. They used their proposed method for the Munich subway tunnel project. Santos et al. presented a procedure to carry out back analysis of the tunnel excavation problem in an automatic manner [27]. They verified their measurements with ones obtained from a finite element analysis. However, in parametric methods all the parameters have physical meaning in the equations and may be used as a method to identify the parameters of the structures. Lozano-Galant et al. proposed a parametric method to uniquely identify the mechanical properties of structures as well as the flexural and axial stiffness (EA, EI) [28]. This technique is called the observability method (OM) and is enforceable for the structural system identification of 1D elements (Bernoulli and Timoshenko beam elements) but not for finite element models with 2D elements.

OM stands as one of the unique structural system identification parametric approaches presented in the literature for the static [29–36] and dynamic analysis of the stiffness matrix method [37]. The advantages of using this deterministic and physics-based approach are: 1—the mathematical foundation of the technique is simple and comprehensible; 2—OM defines whether all the variables or a subset of them are observable or not; 3—OM permits engineers to acquire unique solutions of a structural system in a reasonable time; 4—OM allows the identification of the mechanical parameters of the structure even if the involved parameters are not linearly related; 5—if there is not a unique solution for a specific part of a system, with a quick evaluation of the system relations are obtained that allow the user to specify which measurements must be obtained to have more or complete observability [38].

OM was applied satisfactorily to solve engineering structural problems. For instance, Lozano-Galant et al. facilitated the application of OM for the identification of the parameters of complex concrete and steel bridges by proposing a graphical method for the selection of measurement sets, which is called observability trees [34]. The use of observability trees proved that, in the analyzed system, the probability of selecting the appropriate measurement set with a minimum number of measurements at random was practically negligible. In addition, Castillo et al. utilized OM to assist in decision-making and risk management processes during the maintenance and service life of structures [30].

Although it has applicability to many structures, OM has never been applied to models made out of 2D elements. Along with the undertaken studies and the proposed existing gaps, for the first time we propose a new system of equation for the application of OM to the inverse analysis of 2D models. This application is not as straightforward as it might look. The 1D OM unknowns (Young's modulus E, area A, Inertia I, and flexural or axial stiffnesses) are located only in the numerator of the stiffness matrix [28]. However, for the implementation of 2D models an obstacle arises, as the Poisson ratio appears in the both the numerator and the denominator of the elements of the stiffness matrix. This location of unknowns creates a nonlinear system of equations that prevents defining the unknowns through traditional OM. The location of the unknowns creates a non-linear system of equations that prevents the identification of the mechanical properties of the 2D element structures with the traditional OM. To solve this problem, this paper proposes a new methodology to enable the application of the OM to 2D model structures. The main contribution of this work refers to the use of changes in variables to linearize non-linear systems of equations. To illustrate the introduction of this change of variable into the observability method, a detailed mathematical analysis is presented. Finally, to validate the applicability of the method, the mechanical properties of a state-of-the-art plate are numerically determined. It is important to highlight that the proposed mathematical methodology is characterized by its generality, as it can be used for other systems of equations, including unknown variables in both the numerator and the denominator. With the proposed methodology, scholars can derive the mechanical parameters of crucial structures, such as tunnels, gravity dams, retaining walls, buttress dams, culverts, and underground pipelines.

This paper is organized as follows. In Section 2, the observability technique is elucidated and we demonstrate how it could be exerted to the stiffness matrix method of 2D elements. In Section 3, an example of a state-of-the-art technique is analyzed to show the applicability of OM to 2D element structures. Moreover, a statistical analysis of the required number of measurements for reaching different levels of observability is shown in this section. Finally, the conclusions of the article are presented in Section 4.

2. Structural System Identification of Structures with 2D Elements by Observability Techniques

In this section, the application of the observability method (OM) to the inverse analysis of structures modeled by 2D elements is firstly described. Then, the proposed algorithm to implement the application of the OM to 2D element structures is presented. Finally, the differences in the system of equations of structures with 1D and 2D elements are reviewed.

2.1. Inverse Analysis of the Stiffness Matrix Method

According to the Finite Element Method ([39]), the stiffness matrix of 2D element structures [K] can be obtained from the strain-displacement matrix [B] and elastic matrix [D], as presented in Equation (1):

$$[K] = \int_V [B]^T [D][B] \, dV. \tag{1}$$

In the stiffness matrix method, matrix [K] is used to relate the equilibrium equations in terms of the nodal displacements, as presented in Equation (2).

$$[K]^{(2N_N \times 2N_N)} \cdot \{\delta\}^{(2N_N \times 1)} = \{f\}^{(2N_N \times 1)}, \tag{2}$$

where, [K] depends on the following characteristics: thickness (h), Young's modulus (E), and Poisson ratio (ϑ). $\{\delta\}$ is the vector of displacements, which includes the vertical (v_i) and horizontal (u_i) deflections at each node i, and $\{f\}$ is the force vector that contains the vertical (V_i) and horizontal (H_i) applied forces at each node i. Finally, N_N refers to the number of nodes in the finite element equation. For the case of tunnel, dam, and culvert analysis, the plane strain assumption is traditionally considered and the thickness is assumed as one.

In the traditional application of the stiffness matrix method, the parameters in [K] are known and the nodal displacements can be directly obtained from Equation (2). Then, the internal forces in the elements as well as the reactions at the boundary conditions can be obtained from these displacements. This methodology is traditionally used by computer software to simulate the structural responses. A detailed explanation of the application of this approach can be found in [38].

The stiffness matrix method can be also used to identify unknown mechanical properties of the structural elements (e.g., h_j, ϑ_j, E_j) from the structural response measured on site (inverse approach). In this approach, some terms in the stiffness matrix are assumed as unknown together with the reactions at the boundary conditions and the unmeasured deflections on site. In this inverse approach, non-linear products of unknown variables appear. To identify these non-linear products of unknown in 1D element structures, the OM was proposed in the literature. This method requires the rearrangement of the system of equations of the stiffness matrix method to join all the unknown variables together. To do so, the system in Equation (2) can be reordered as follows [28]:

$$[K]^* \cdot \{\delta\}^* = \{f\}, \tag{3}$$

where, the products of variables are located in the modified vector of displacements $\{\delta^*\}$ and the modified stiffness matrix $[K]^*$ is a matrix of coefficients with different dimensions than the initial stiffness matrix $[K]$. Once the boundary conditions and the applied forces at the nodes during the nondestructive test are introduced, it can be assumed that a subset of deflections δ_1^* of $\{\delta^*\}$ and a subset of forces in nodes f_1 of $\{f\}$ are known and the remaining subset δ_0^* of $\{\delta^*\}$ and f_0 of $\{f\}$ are unknown. By the static condensation procedure, the system in (3) can be partitioned as follows [28]:

$$[K]^* \cdot \{\delta^*\} = \begin{bmatrix} K_{00}^* & K_{01}^* \\ K_{10}^* & K_{11}^* \end{bmatrix} \cdot \left\{ \begin{array}{c} \delta_0^* \\ \delta_1^* \end{array} \right\} = \left\{ \begin{array}{c} f_0 \\ f_1 \end{array} \right\} = \{f\}, \tag{4}$$

where, K_{00}^*, K_{01}^*, K_{10}^*, and K_{11}^* are the partitioned matrices of $[K]^*$ and δ_0^*, δ_1^*, f_0, and f_1 are the partitioned vectors of $\{\delta^*\}$ and $\{f\}$. In order to join the unknowns, system (4) can be written in the equivalent form [29]:

$$[C] \cdot \{z\} = \begin{bmatrix} K_{10}^* & 0 \\ K_{00}^* & -I \end{bmatrix} \cdot \left\{ \begin{array}{c} \delta_0^* \\ f_0 \end{array} \right\} = \left\{ \begin{array}{c} f_1 - K_{11}^* \delta_1^* \\ K_{01}^* \delta_1^* \end{array} \right\} = \{f^*\}, \tag{5}$$

where, 0 and are the null and the identity matrices, respectively. In this system, the vector of unknown variables, $\{z\}$, appears on the left-hand side and the vector of observations, $\{f^*\}$, on the right-hand side. Both vectors are related by a coefficient matrix $[C]$. A summary of the couple variables and the nonlinear product of unknowns appearing in vector $\{\delta^*\}$ for the case of 2D element structures (modeled by 6-noded triangular element) is presented in Table 1.

Table 1. Coupled variables and products of unknowns in the system of equation.

Coupled variables	$\frac{E_i h_i}{\theta_i + 1}$	$\frac{E_i h_i}{2\theta_i^2 + \theta_i - 1}$	$\frac{E_i h_i (\theta_i - 1)}{2\theta_i^2 + \theta_i - 1}$	$\frac{E_i h_i (5\theta_i - 3)}{2\theta_i^2 + \theta_i - 1}$	$\frac{E_i h_i (10\theta_i - 9)}{2\theta_i^2 + \theta_i - 1}$
Nonlinear products of unknowns	$\frac{E_i h_i}{\theta_i + 1} \cdot u_i$ $\frac{E_i h_i}{\theta_i + 1} \cdot v_i$	$\frac{E_i h_i}{2\theta_i^2 + \theta_i - 1} \cdot u_i$ $\frac{E_i h_i}{2\theta_i^2 + \theta_i - 1} \cdot v_i$	$\frac{E_i h_i (\theta_i - 1)}{2\theta_i^2 + \theta_i - 1} \cdot u_i$ $\frac{E_i h_i (\theta_i - 1)}{2\theta_i^2 + \theta_i - 1} \cdot v_i$	$\frac{E_i h_i (5\theta_i - 3)}{2\theta_i^2 + \theta_i - 1} \cdot u_i$ $\frac{E_i h_i (5\theta_i - 3)}{2\theta_i^2 + \theta_i - 1} \cdot v_i$	$\frac{E_i h_i (10\theta_i - 9)}{2\theta_i^2 + \theta_i - 1} \cdot u_i$ $\frac{E_i h_i (10\theta_i - 9)}{2\theta_i^2 + \theta_i - 1} \cdot v_i$

In order to check if the system has a solution, it is sufficient to calculate the null space $[V]$ of $[C]$ and check that $[V][C] = \{0\}$. The general solution (the set of all solutions) of Equation (5) has the structure of Equation (6) (see Castillo et al. [39]):

$$\{z\} = \left\{ \begin{array}{c} \delta_{00}^* \\ f_{00} \end{array} \right\} + [V] \cdot \{\rho\}, \tag{6}$$

where, $\{z\}$ refers to the general solution and $\left\{ \begin{array}{c} \delta_{00}^* \\ f_{00} \end{array} \right\}$ is a particular solution of the system (6). This particular solution may be derived by different subroutines in Matlab [40] either parametrically (backslash function ($\backslash$)) or numerically (Moore–Penrose pseudoinverse function (pinv)). The product $[V]\{\rho\}$ represents the set of all solutions of the associated homogeneous system of equations (a linear space of solutions, where the columns of $[V]$ are basis and the elements of the column matrix, and $\{\rho\}$ are arbitrary real values which represent the coefficients of all possible linear combinations). It is interesting to note that a variable has a unique solution not only when matrix $[V]$ has zero dimensions but also when the associated row in matrix $[V]$ is null. This matrix might provide full (FO) or partial (PO) observability, depending on the number of parameters identified. The PO might help to identify new parameters in the structure through a recursive process. In this recursive process, the observed information is successively introduced as input data in the observability analysis. For a more detailed analysis of the mathematical manipulation of the system of equations presented, the reader is addressed to Lozano-Galant et al. [28]. In this work, detailed step by step analyses of the application of the observability techniques to beam structures (1D element) are presented.

The application of the observability techniques to the identification of 2D element structures is not presented in the literature. In this work, triangular elements are assumed. This type of element has been extensively used in the literature ([41]). In fact, this geometry does not include some of the limitations of the straight-sided elements, which displacement functions induce their vertices to stay straight without allowing curved deformations. In addition, they present a higher convergence rate. Three-noded constant strain triangle (CST) elements can be satisfactorily used for meshing complex structures, as they have lower accuracy than linear strain triangle (LST) ones. Although some scholars have improved both the accuracy and convergency of CST elements (see, e.g., Yang et al. [42], Piltner and Taylor [43], and Neto et al. [44]), in this work LST elements are considered.

Although how it may appear, the application of OM to the structures modeled by LST elements is not straightforward, as ϑ is located in both the numerator and denominator of the stiffness matrix parameters. It can be seen in Table 1 that this system of equations is non-linear, as coupled variables are coupled to other unknowns (nodal displacements). In order to solve the above-cited issue and linearize the system of equations, the following change in variables is proposed:

$$NU_i^p = 1/(\vartheta_i + 1), \tag{7}$$

$$NU_i^N = 1/(2\vartheta_i - 1). \tag{8}$$

The super indices in these equations refer to the positive (P) and negative (N) signs in the denominator. Table 2 presents the parameters that may appear in the classical stiffness matrix of the structures modeled by LST elements, the above change in variables, and the updated variables in the system of equations.

Table 2. Unknown variables before and after the proposed changes in variables.

Unknowns	Change of Variables	Updated Unknowns
$\frac{E_i h_i}{\vartheta_i + 1}$		$E_i.h_i.NU_i^p$
$\frac{E_i h_i}{2\vartheta_i^2 + \vartheta_i - 1}$		$E_i.h_i.NU_i^p.NU_i^N$
$\frac{E_i h_i \vartheta_i}{2\vartheta_i^2 + \vartheta_i - 1}$	$NU_i^p = \frac{1}{\vartheta_i + 1}$	$E_i.h_i.\vartheta_i.NU_i^p.NU_i^N$
$\frac{E_i h_i (\vartheta_i - 1)}{2\vartheta_i^2 + \vartheta_i - 1}$	$NU_i^N = \frac{1}{2\vartheta_i - 1}$	$E_i.h_i.\vartheta_i.NU_i^p.NU_i^N - E_i.h_i.NU_i^p.NU_i^N$
$\frac{E_i h_i (5\vartheta_i - 3)}{2\vartheta_i^2 + \vartheta_i - 1}$		$5.E_i.h_i.\vartheta_i.NU_i^p.NU_i^N - 3.E_i.h_i.NU_i^p.NU_i^N$
$\frac{E_i h_i (10\vartheta_i - 9)}{2\vartheta_i^2 + \vartheta_i - 1}$		$10.E_i.h_i.\vartheta_i.NU_i^p.NU_i^N - 9.E_i.h_i.NU_i^p.NU_i^N$

As shown in Table 2, after the proposed change in variables in $[K]^*$, the unknown variables are generated and reduced to the following products: $\left\{E_i.h_i.NU_i^p\right\}$, $\left\{E_i.h_i.NU_i^p.NU_i^N\right\}$, and $\left\{E_i.h_i.\vartheta_i.NU_i^p.NU_i^N\right\}$. This paper is focused on plane strain structures, where h_i is assumed to be 1. In comparison with the traditional observability techniques, dealing with this type of structure produces the following difficulties: (1) The product of unknown variables (nonlinearity of the problem): The following variables (such as E_i, ϑ_i, NU_i^p, and NU_i^N) are unknown and coupled to each other. To solve this issue, the target unknowns are identified as monomial-coupled elements (for example, the unknown variable $E_i.h_i.NU_i^p$ is considered as a new linear variable named $E_i h_i NU_i^p$). In addition, the observed products of variables can be used to calculate the value of single variables (e.g., using E_i to uncouple $\{E_i.\vartheta_i\}$). (2) The components of the stiffness matrix might be composed of the sum of different linearized products of variables (see, for example, the last three rows of Table 2).

To deal with this problem, the columns of [K]* were split, following the methodology in the traditional application of OM [28]. This operation increases the size of Equation (3) to:

$$[K]^{*(2N_N \times (2N_N \times E_N \times 3))} \cdot \{\delta\}^{*((2N_N \times E_N \times 3) \times 1)} = \{f\}^{(2N_N \times 1)}, \qquad (9)$$

where, N_N refers to the number of nodes and E_N to the number of elements. (3) The calculation of the null space. The application of OM to the 1D element structures enables the symbolical calculation of the null space. This symbolical approach enables the visualization of the dependences on the variables. Unfortunately, the symbolical calculation of the null space in 2D element structures is very time-consuming; thus, a numerical calculation was used for this research. This numerical analysis can reduce the computation time by more than 99.6%.

2.2. Proposed Algorithm

To illustrate the application of the procedure presented above, an algorithm is proposed in this section for the structural system identification of 2D element structures with the OM. This algorithm is presented in Figure 1. The whole package of the 2D observability technique was written in MatLab [40]. This package contains two main connected parts. The first part contains only direct analysis codes. In this part, 2D finite element analyses were developed for the plane strain analysis and the user only needs to input the node coordinates, the mechanical properties, the boundary conditions, and the applied forces. Afterwards, through the interface the user selects the defined structure and runs the direct analysis to calculate the deflections at the free nodes and the reactions at the boundary conditions. Some of the information calculated in this step is used as input data for the inverse analysis of the structure in the second part. The second part of the package was only assigned to inverse analysis and contains the procedure presented in the preceding section. To carry out the inverse analysis, the user has to define the condition of the inverse analysis through the interface. In this step, the interface offers all the possible measurements of the structure calculated in the direct analysis. After selecting the measured degrees of freedom, an observability analysis is performed and the observable variables are provided together with their numerical values. The steps of the algorithm for the inverse analysis of the structure are as follows:

- Step 1. Generate the classical stiffness matrix equation presented in Equation (2).
- Step 2. Exert the change in variable to the stiffness matrix using the functions of Poisson's ratio presented in Equations (7) and (8).
- Step 3. Establish Equation (3) considering the variables defined in Step 2.
- Step 4. Reorder the columns in matrix to isolate the monomial products of variables.
- Step 5. Introduce the boundary conditions, the known forces, and the measured deflections.
- Step 6. Reorder the system following Equation (5).
- Step 7. Calculate the null space of [C] numerically and identify the observed variables.
- Step 8. Calculate the particular solution of the system numerically using the Moore–Penrose pseudoinverse function.
- Step 9. Use the observed parameters or the observed coupled variables to simplify the other coupled variables.
- Step 10. Introduce the observed parameters into Step 5. Repeat until no additional parameters are observed (end of the recursive process).

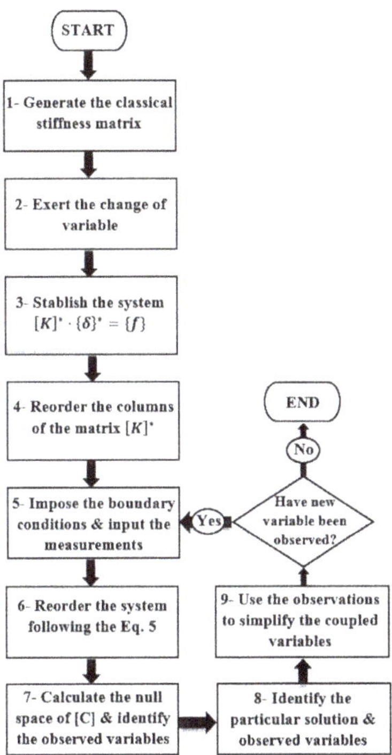

Figure 1. Flow chart of the algorithm.

A summary of the aforementioned algorithm is illustrated in Figure 1.

2.3. Comparison of the Application of OM to the 1D and 2D Element Structures

To illustrate the effect of changing the type of element into the observability method, Table 3 is presented. This table compares the application of 1D (2-noded beam element) and 2D (6-noded triangular element, LST) element structures and includes the following information: (1) type of element, (2) degrees of freedom per element, (3) unknown variables, (4) products of unknown variables, and (5) size of the stiffness matrix and, (6) type of analysis of the null space. The information of the 2D elements is presented before and after the change in variables described in Section 1.

Table 3. Comparison of 1D and 2D observability methods (OMs).

	1D OM	2D OM	
		Before Change of Var.	After Change of Var.
Type of element	2-noded beam element	6-noded triangular element	6-noded triangular element
Degrees of freedom per element	6	12	12
Number of unknows	2	2	4
Unknown variables per element	$E_i A_i$, $E_i I_i$	E_i and ϑ_i	E_i, ϑ_i, NU_i^p, NU_i^N
Products of unknowns	$(E_i A_i / L_i)$ *and* $(E_i I_i / L_i)$	$(E_i h_i / \vartheta_i + 1)$ $(E_i h_i / 2\vartheta_i^2 + \vartheta_i - 1)$, $(E_i h_i \vartheta_i / 2\vartheta_i^2 + \vartheta_i - 1)$	$E_i . h_i . NU_i^p$ $E_i . h_i . NU_i^p . NU_i^N$, $E_i . h_i . \vartheta_i . NU_i^p . NU_i^N$
Size of stiffness matrix	$[3N_N \times 3N_N]$	$[2N_N \times 2N_N]$	$[2N_N \times (2N_N \times E_N \times 3)]$
Calculation of the null space	Symbolical	Numerical	Numerical

The analysis of Table 3 shows that the number of unknown variables per element is doubled (changing from 2 to 4) after applying the change of variables in the 2D elements. Therefore, the number of columns of the stiffness matrix is also consequently increased. This table also illustrates that, unlike the analysis of 1D elements, a numerical simulation of the null space is required for the analysis of 2D element structures.

3. Example of the Application

In this section, a state-of-the-art example is analyzed to illustrate the applicability of the proposed algorithm to the structural system identification of 2D element structures with observability techniques. Firstly, the geometry, boundary, and loading conditions of the example are detailed together with the results of the direct analysis. Then, some steps of the inverse analysis of the structure are detailed. Finally, a statistical analysis is presented to illustrate the role of the measuring set on the observability of the structural parameters of this structure.

3.1. Definition of the Analyzed Structure

The analyzed structure was chosen from Peter Kattan's book [45]. This structure corresponds with the discretization of a clamped plate simulated with the plane strain theory. For the finite element model, two quadratic 6-noded LST elements (see Figure 2a) were considered, leading to a system with 9 nodes and 18 degrees of freedom. The boundary and loading conditions as well as the node and element numbering are presented in Figure 2b. The uniform force at the right edge of the structure (corresponding to a distributed load of W = 3000 kN/m^2) has been simulated as concentrated forces at the element nodes (nodes 3, 6, and 9), according to the tributary area of each node. As illustrated in this figure, the clamp conditions are modeled by fixing the horizontal and vertical deflections of nodes 1, 4, and 7. The horizontal and vertical deflections of the rest of the nodes in the structure remain free.

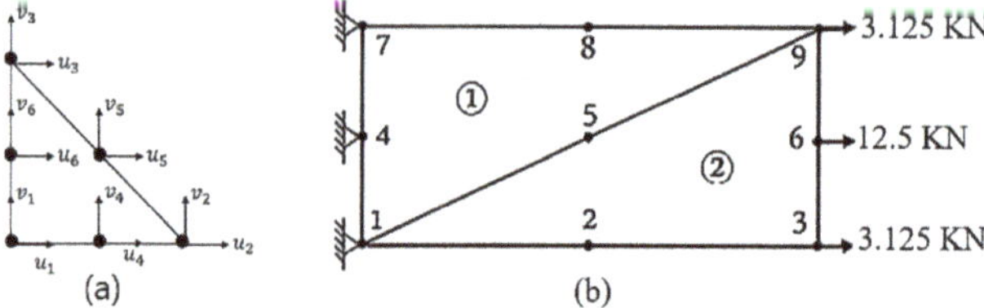

Figure 2. Example: (**a**) 6-noded triangular element (LST), and (**b**) node and element numbering together with the boundary and loading conditions [45].

The analyzed structure has a length of 0.5 m, a height of 0.25 m, and the thickness of 1 (plane strain assumption). The Young's modulus and Poisson's ratio are E = 210 GPa and ϑ = 0.3, respectively. The first part of the developed package was used to calculate the horizontal and vertical deflections at the free nodes as well as the reactions at the boundary conditions (direct analysis). The obtained deflections and reactions are summarized in Table 4. The results obtained from the direct analysis are considered in good agreement with those in [45] (in fact, both results present the same number of decimal numbers).

Table 4. Deflections and reactions at the example from the direct analysis.

Node	1	2	3	4	5	6	7	8	9
u_i(m) e-6		0.080	0.1580		0.0739	0.1568		0.0716	0.1580
v_i(m) e-6		0.0227	0.0288		0.0055	0.0113		−0.011	−0.007
H_i(KN)	−3.541			−11.67			−3.541		
V_i(KN)	−2.863			0.012			2.851		

3.2. Inverse Analysis

In this section, the proposed algorithm is used for the inverse analysis of the structure from the following measurement set (u_2, v_2, u_3, v_3, u_5, v_5, u_6, v_6, u_9, and v_9). The numerical values of these deflections correspond with those presented in Table 4.

To identify the observable parameters in the structure, the stiffness matrix system described in Step 1 of the algorithm was calculated. This example contains 18 components for displacements (9 for both u and v). Unfortunately, the large size of this matrix (18×18) prevents its representation in the paper and only the first equation (the one related to the horizontal force at node 1, H_1 is depicted) is presented in Equation (10):

$$H_1 = \begin{bmatrix} ((E_1.h_1)/2(\vartheta_1+1)) + ((E_2.h_2)/(\vartheta_2-1))/((4.(2\vartheta_2-1).(\vartheta_2+1)) \\ 0 \\ ((E_2.h_2).(\vartheta_2-1))/(3.(2\vartheta_2-1).(\vartheta_2+1)) \\ -(2.E_2.h_2.\vartheta_2)/(3(2.\vartheta_2-1).(\vartheta_2+1)) \\ (E_2.h_2.(\vartheta_2-1))/(12(2\vartheta_2-1).(\vartheta_2+1)) \\ (E_2.h_2.\vartheta_2)/6.(2\vartheta_2-1).(\vartheta_2+1) \\ -(2.E_1.h_1)/3.(\vartheta_1+1) \\ (E_1.h_1)/3.(\vartheta_1+1) \\ 0 \\ ((2.E_2.h_2.\vartheta_2)/(3(2\vartheta_2-1).(\vartheta_2+1))) - ((E_1.h_1)/3(\vartheta_1+1)) \\ 0 \\ 0 \\ (E_1.h_1)/(6(\vartheta_1+1)) \\ -(E_1.h_1)/(12(\vartheta_1+1)) \\ 0 \\ 0 \\ 0 \\ ((E_1.h_1)/2(\vartheta_1+1)) - ((E_2.h_2.\vartheta_2)/(6(2\vartheta_2-1).(\vartheta_2+1))) \end{bmatrix}^T \cdot \begin{Bmatrix} u_1 \\ v_1 \\ u_2 \\ v_2 \\ u_3 \\ v_3 \\ u_4 \\ v_4 \\ u_5 \\ v_5 \\ u_6 \\ v_6 \\ u_7 \\ v_7 \\ u_8 \\ v_8 \\ u_9 \\ v_9 \end{Bmatrix} \tag{10}$$

As shown in Equation (10), Poisson's ratio appears in both the numerator and denominator of the matrix parameters. This location of Poisson's ratio prevents the application of traditional OM (proposed in the literature) and makes necessary the changes in variables proposed in Step 2. After this change in variables, the system of equations can be rearranged. In this way, the size of the modified stiffness matrix is increased to 18×42. Again, for size limitations only the obtained equilibrium equation of H_1 is presented in the paper. This equation is as follows:

$$
H_1 = \begin{bmatrix}
0 \\
0 \\
0 \\
0 \\
0 \\
0 \\
-1/3 \\
1/2 \\
0 \\
0 \\
-2/3 \\
1/6 \\
2/3 \\
0 \\
-1/6 \\
0 \\
0 \\
0 \\
0 \\
0 \\
0 \\
0 \\
-1/3 \\
0 \\
-1/12 \\
-1/3 \\
-1/12 \\
0 \\
0 \\
0 \\
0 \\
0 \\
0 \\
0 \\
0 \\
0 \\
0 \\
0 \\
0 \\
0 \\
0
\end{bmatrix}^{T} \cdot
\left\{
\begin{array}{l}
E_1.\vartheta_1.NU_1^{p}.NU_1^{N}.u_5 \\
E_1.\vartheta_1.NU_1^{p}.NU_1^{N}.u_8 \\
E_1.\vartheta_1.NU_1^{p}.NU_1^{N}.u_9 \\
E_1.\vartheta_1.NU_1^{p}.NU_1^{N}.v_5 \\
E_1.\vartheta_1.NU_1^{p}.NU_1^{N}.v_8 \\
E_2.\vartheta_2.NU_2^{p}.NU_2^{N}.u_2 \\
E_2.\vartheta_2.NU_2^{p}.NU_2^{N}.u_3 \\
E_2.\vartheta_2.NU_2^{p}.NU_2^{N}.u_5 \\
E_2.\vartheta_2.NU_2^{p}.NU_2^{N}.u_6 \\
E_2.\vartheta_2.NU_2^{p}.NU_2^{N}.v_2 \\
E_2.\vartheta_2.NU_2^{p}.NU_2^{N}.v_3 \\
E_2.\vartheta_2.NU_2^{p}.NU_2^{N}.v_5 \\
E_2.\vartheta_2.NU_2^{p}.NU_2^{N}.v_6 \\
E_2.\vartheta_2.NU_2^{p}.NU_2^{N}.v_9 \\
E_1.NU_1^{p}.NU_1^{N}.u_5 \\
E_1.NU_1^{p}.NU_1^{N}.u_8 \\
E_1.NU_1^{p}.NU_1^{N}.u_9 \\
E_1.NU_1^{p}.NU_1^{N}.v_5 \\
E_1.NU_1^{p}.NU_1^{N}.v_8 \\
E_1.NU_1^{p}.u_5 \\
E_1.NU_1^{p}.u_8 \\
E_1.NU_1^{p}.v_5 \\
E_1.NU_1^{p}.v_8 \\
E_1.NU_1^{p}.v_9 \\
E_2.NU_2^{p}.NU_2^{N}.u_2 \\
E_2.NU_2^{p}.NU_2^{N}.u_3 \\
E_2.NU_2^{p}.NU_2^{N}.u_5 \\
E_2.NU_2^{p}.NU_2^{N}.u_6 \\
E_2.NU_2^{p}.NU_2^{N}.v_2 \\
E_2.NU_2^{p}.NU_2^{N}.v_3 \\
E_2.NU_2^{p}.NU_2^{N}.v_5 \\
E_2.NU_2^{p}.NU_2^{N}.v_6 \\
E_2.NU_2^{p}.NU_2^{N}.v_9 \\
E_2.NU_2^{p}.u_2 \\
E_2.NU_2^{p}.u_3 \\
E_2.NU_2^{p}.u_5 \\
E_2.NU_2^{p}.u_6 \\
E_2.NU_2^{p}.u_9 \\
E_2.NU_2^{p}.v_2 \\
E_2.NU_2^{p}.v_3 \\
E_2.NU_2^{p}.v_5 \\
E_2.NU_2^{p}.v_6
\end{array}
\right\}
\tag{11}
$$

After introducing the known information (null deflections at the boundary conditions; deflections at the measurement set; and known forces at nodes 3, 6 and 9), the system of equations was arranged as presented in Step 6. To proceed this application, the equilibrium equation of the horizontal reaction at node 2, H_2, is presented. This equation is as follows:

$$
H_2 = \begin{bmatrix} 0 \\ 0 \\ 0 \\ 2 \times 10^{-07} \\ 0 \\ 0 \\ 0 \\ 0 \\ 0 \\ 0 \\ 9 \times 10^{-08} \\ 9 \times 10^{-08} \\ 0 \\ 0 \\ 0 \\ 0 \\ 0 \\ 0 \end{bmatrix}^T \cdot \left\{ \begin{array}{c} E_1.\vartheta_1.NU_1^p.NU_1^N \\ E_1.\vartheta_1.NU_1^p.NU_1^N.u_8 \\ E_1.\vartheta_1.NU_1^p.NU_1^N.v_8 \\ E_2.\vartheta_2.NU_2^p.NU_2^N \\ E_1.NU_1^p \\ E_1.NU_1^p.NU_1^N \\ E_1.NU_1^p.NU_1^N.u_8 \\ E_1.NU_1^p.NU_1^N.v_8 \\ E_1.NU_1^p.u_8 \\ E_1.NU_1^p.v_8 \\ E_2.NU_2^p \\ E_2.NU_2^p.NU_2^N \\ H_1 \\ H_4 \\ H_7 \\ V_1 \\ V_4 \\ V_7 \end{array} \right\}
\tag{12}
$$

This equation shows that at this stage, the number of unknown variables in the system is eight (E_1, ϑ_1, NU_1^P, NU_1^N, E_2, ϑ_2, NU_2^P, NU_2^N). To identify the observability of the variables, the null space [V] was numerically calculated in Step 7. With this information, the general solution of the system {z} described in Equation (6) can be written as presented in Figure 3.

$$
\left\{ \begin{array}{c} E_1.\vartheta_1.NU_1^P.NU_1^N \\ E_1.\vartheta_1.NU_1^P.NU_1^N.u_8 \\ E_1.\vartheta_1.NU_1^P.NU_1^N.v_8 \\ E_2.\vartheta_2.NU_2^P.NU_2^N \\ E_1.NU_1^P \\ E_1.NU_1^P.NU_1^N \\ E_1.NU_1^P.NU_1^N.u_8 \\ E_1.NU_1^P.NU_1^N.v_8 \\ E_1.NU_1^P.u_8 \\ E_1.NU_1^P.v_8 \\ E_2.NU_2^P \\ E_2.NU_2^P.NU_2^N \\ H_1 \\ H_4 \\ H_7 \\ V_1 \\ V_4 \\ V_7 \end{array} \right\} = \left\{ \begin{array}{c} 6.9 \times 10^{+04} \\ 2.4 \times 10^{+03} \\ 3.8 \times 10^{+02} \\ -1.2 \times 10^{+08} \\ 1.5 \times 10^{10} \\ 2.2 \times 10^{+09} \\ 3.5 \times 10^{+03} \\ 8.2 \times 10^{+02} \\ -4.3 \times 10^{+02} \\ -3.2 \times 10^{+03} \\ 1.6 \times 10^{+08} \\ -4.0 \times 10^{+08} \\ -4.0 \times 10^{01} \\ 3.8 \times 10^{+02} \\ -7.4 \times 10^{+02} \\ -8.3 \times 10^{-02} \\ 1.0 \times 10^{+03} \\ -1.1 \times 10^{+03} \end{array} \right\} + \left(\begin{array}{ccc} 0.4 \times 10^{+08} & 0 & 0 \\ -3 \times 10^0 & -8 \times 10^0 & -9 \times 10^0 \\ 8 \times 10^{-2} & 2 \times 10^0 & 2e \times 10^0 \\ 0 & 0 & 0 \\ -2.5 \times 10^{+08} & -3.5 \times 10^{+08} & -3.6 \times 10^{+08} \\ 1.6 \times 10^{+08} & 0.9 \times 10^{+08} & 0.8 \times 10^{+08} \\ 1 \times 10^{01} & 7 \times 10^0 & 4 \times 10^0 \\ -1 \times 10^0 & 3 \times 10^0 & 2 \times 10^0 \\ -9 \times 10^0 & -1 \times 10^0 & -1 \times 10^0 \\ 1 \times 10^{-07} & 3 \times 10^{-07} & 3 \times 10^{-08} \\ 0 & 0 & 0 \\ 0 & 0 & 0 \\ 6 \times 10^{-1} & 8 \times 10^{-01} & 8 \times 10^{-01} \\ 5 \times 10^0 & 5 \times 10^0 & 4 \times 10^0 \\ 9 \times 10^0 & 1 \times 10^1 & 1 \times 10^1 \\ 1 \times 10^0 & 0 & 0 \\ 0 & 1 \times 10^0 & 0 \\ 0 & 0 & 1 \times 10^0 \end{array} \right) \cdot \left\{ \begin{array}{c} \rho_1 \\ \rho_2 \\ \rho_3 \end{array} \right\}
$$

Figure 3. General solution of the system of equation.

The analysis of the null space presented in Figure 3 shows that the partial observability of the structure is obtained. The particular solution of these parameters was numerically obtained with the pseudoinverse of matrix [C] in MatLab (Step 8). In Figure 3, the observable parameters (the ones with a unique solution) are associated with the null rows of the matrix [V]. The values of these parameters are $(\vartheta_2.E_2.NU_2^P.NU_2^N) = -1.2 \times 10^{+08}$, $(E_2.NU_2^P. NU_2^N) = -4.0 \times 10^{+08}$, and $(E_2.NU_2^P) = 1.6 \times 10^{+08}$. Finally, by dividing the observable coupled variables to each other, the following results were obtained for element 2 of the structure: $NU_2^P = 7.7 \times 10^{-01}$, $NU_2^N = -25 \times 10^{-1}$, $E_2 = 2.1 \times 10^{+08}$ Pa, and $\vartheta_2 = 3 \times 10^{-01}$. Increasing the number of measurements may lead to increasing the number of observed parameters or full observability. To illustrate the effect of the number of measurements on the observability of the system, a statistical analysis is presented in the following section.

3.3. Statistical Analysis

In this part, a statistical analysis of the example is presented to indicate the required number of measurements for the partial and full observability of the system. To do so, combinatorial analysis was conducted and the following combinatory equation was used:

$$C(n, r) = (n!)/(r!(n - r)!), \tag{13}$$

with n as the number of possible measurement and r as the chosen number of measurements in each combination. In the analyzed example, n is equal to 12, as the possible measurements correspond to u_2, v_2, u_3, v_3, u_5, v_5, u_6, v_6, u_8, v_8, u_9, and v_9. This combinatorial analysis was carried out for 8 ($C_{12}^8 = 495$), 9 ($C_{12}^9 = 220$), 10 ($C_{12}^{10} = 66$), 11 ($C_{12}^{11} = 12$), and 12 ($C_{12}^{12} = 1$) measurements per set. It can be seen in Figure 4 that the full observability of the structure requires at least nine measurements per set. In the case of eight measurements, 0.6% of the sets lead to the partial observability of the system, and 99.4% lead to an indeterminate system. For the case of nine measurements, 3.6% lead to the full observability of the system, 4.6% of the sets lead to the partial observability of the system, and 91.8% lead to an indeterminate system. This figure also shows that by conducting 10 measurements, 72.6% of the measurements lead to the full observability of the system, 1.5% of the sets lead to the partial observability of the system, and 25.9% lead to an indeterminate system.

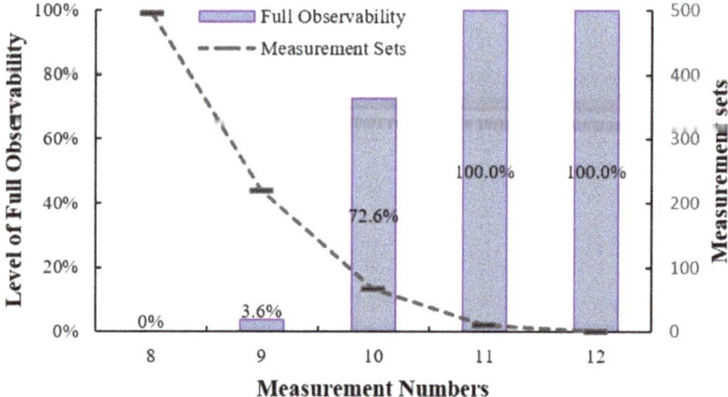

Figure 4. Full observability of the system based on the various measurement sets.

As illustrated in Table 5, to distinguish the difference between the sets achieving different levels of observability, they were classified into different patterns with regard to the location of the measurements. For instance, in the case of eight measurements $\{u|v_2, u_3, v_3, u_5, v_5, u_6, v_6, u|v_9\}$, $\{u_2, v_2, u_3, v_3, u_5, v_5, u|v_6, u|v_9\}$, and $\{u_2, v_2, u_3, v_3, u_5, v_5, u_6, v_6\}$ were classified as $\{2, \overline{3}, \overline{5}, \overline{6}, 9\}$, $\{\overline{2}, \overline{3}, \overline{5}, 6, 9\}$, and $\{\overline{2}, \overline{3}, \overline{5}, \overline{6}\}$, respectively (the vertical bar between u and v indicates that either the vertical or horizontal degree of freedom is measured). In these patterns, the subscripts of the measurements (node number) that identify the associated location are shown.

Table 5. All the sets for achieving different levels of observability for 8, 9, and 10 measurements per set. Partial observability (P.O.) and full observability (F.O.).

8 Measurements	9 Measurements		10 Measurements	
P.O.	F.O.	P.O.	F.O.	P.O.
$2, \overline{3}, \overline{5}, \overline{6},9$	$2,\overline{3},\overline{5}, \overline{6},8,9$	$2,\overline{3},\overline{5}, \overline{6}, \overline{9}$	$\overline{3},\overline{5},\overline{6},\overline{8},\overline{9}$	$2, \overline{3}, \overline{5}, \overline{6}, \overline{9}$
$\overline{2}, \overline{3}, \overline{5},6,9$	$2,\overline{5}, \overline{6}, \overline{8}, \overline{9}$	$\overline{2},3,\overline{5}, \overline{6}, \overline{9}$	$2,3,\overline{5}, \overline{6},\overline{8},\overline{9}$	
$\overline{2}, \overline{3}, \overline{5}, \overline{6}$	$\overline{2}, \overline{3}, 5,6,8,9$	$\overline{2},\overline{3},5,\overline{6},\overline{9}$	$2,\overline{3}, \overline{5},6, \overline{8}, \overline{9}$	
	$\overline{2}, \overline{3}, \overline{5}, \overline{6},8$	$\overline{2}, \overline{3}, \overline{5},6,\overline{9}$	$2, \overline{3}, \overline{5}, \overline{6},8, \overline{9}$	
		$\overline{2}, \overline{3}, \overline{5},6,9$	$2, \overline{3}, \overline{5}, \overline{6}, \overline{8},9$	
			$\overline{2}, \overline{5}, \overline{6}, \overline{8}, \overline{9}$	
			$\overline{2},, 6,\overline{8}, \overline{9}$	
			$\overline{2},3, \overline{5},\overline{6},8, \overline{9}$	
			$\overline{2},3, \overline{5}, \overline{6}, \overline{8},9$	
			$\overline{2}, \overline{3},5,\overline{6},8, \overline{9}$	
			$\overline{2}, \overline{3},\overline{5},\overline{8}, \overline{9}$	
			$\overline{2}, \overline{3}, \overline{5},6,8, \overline{9}$	
			$\overline{2}, \overline{3}, \overline{5},6, \overline{8},9$	
			$\overline{2}, \overline{3}, \overline{5}, \overline{6}, \overline{8}$	

The node numbers with the sign $(-)$ indicate that both deflections, vertical and horizontal, are measurements. However, no distinction is made between the measurements of X and Y degrees of freedom, as the ones without any sign are indicative of either vertical or horizontal measurements. All the sets related to the PO of the system in the cases of 8, 9, and 10 measurements are listed in columns 1, 3, and 5 of Table 5, respectively. However, columns 2 and 4 include the sets for FO of the system based on 9 and 10 measurements. One of the surprising results of the method is that, with eight targeted unknowns, at least nine measurements are required to obtain full observability. This is one of the drawbacks inherent in the method. As the nonlinear problem is being solved as a linear one and coupled unknowns are linearized, extra information might be required to uncouple them and to obtain full observability.

It is clear that no set with 8, 9, or 10 measurements leads to the FO of the examined structure. Hence, if those are not elected appropriately, the end of the recursive process occurs before identifying all variables. In Table 5, the sets leading to the FO of the system are selected at dispersed locations of the structure. With these sets, in the other words, the distributed placement of the sensors keeps the observability flow, and, consequently, FO will be obtained. If the measurements are conducted intensively at a local area or specific element of the structure, the redundancy of the measurements will emerge.

In the same manner, in the case of $\left\{2,\overline{3},\overline{5}, \overline{6},8,9\right\}$ (the first set related with nine measurements), achieving the FO was due to the dispersed locations of the sensors at all nodes of the structure. Accordingly, for the same number of measurements as well as 10, the occurrence of PO in all sets (third and fifth columns of Table 5) is due to a lack of any information about node 8. For instance, choosing the measurement set, $\left\{\overline{2}, \overline{3}, \overline{5}, \overline{6}, \overline{9}\right\}$ leads to partial observability despite 10 measurements being included in the set.

This set will only allow a unique solution for the following coupled unknowns $\left\{E_2.\vartheta_2.NU_2^p.NU_2^N\right\}$, $\left\{E_2.NU_2^p.NU_2^N\right\}$, and $\left\{E_2.NU_2^p\right\}$. However, the first set in the fourth column of Table 5, $\left\{\overline{3},\overline{5},\overline{6},\overline{8},\overline{9}\right\}$, without any information about node 2 enabled the FO of the system. In the first recursive step of the inverse analysis, the particular solutions for the couple $\left\{E_1.NU_1^p.NU_1^N\right\}$ and $\left\{E_1.NU_1^p\right\}$ related to element 1 as well as $\left\{E_2.\vartheta_2.NU_2^p.NU_2^N\right\}$ and $\left\{E_2.NU_2^p.NU_2^N\right\}$ related to element 2 are acquired. This information is sufficient to derive the value of the parameters NU_1^N and ϑ_2 first and the rest of unknowns of the system. All the above information implies that the reason for the partial observability is that the number of effective measurements is less than the number of unknowns.

Figure 5 illustrates some of the mentioned possible sets (in Table 5) containing both vertical and horizontal measurements. Figure 5a–c presents the sets that lead to the PO of the system by measuring 8, 9, and 10 deflections, respectively. The FO of the problem through 9 and 10 measurements is depicted in Figure 5b1,c1, respectively.

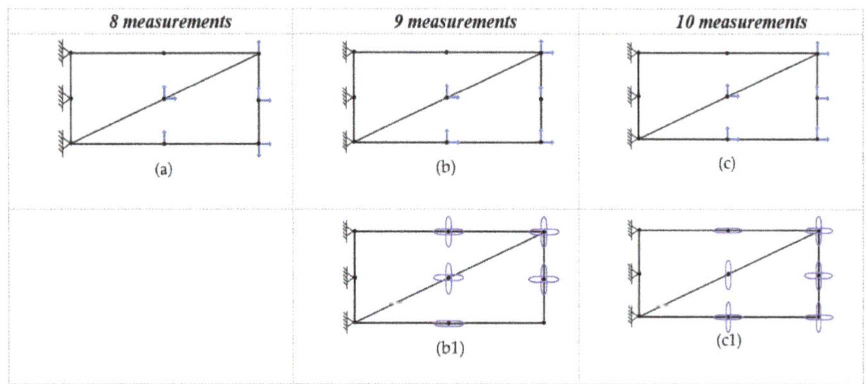

Figure 5. Required number of measurements for having the partial (**a**,**b**,**c**) and full observability (**b1**,**c1**) of the system.

4. Conclusions

The observability technique has been studied in the literature for the structural system identification of 1D element structures (such as beams and trusses). Nevertheless, the application of this method to 2D element structures (such as dams, tunnels, and culverts) has not been studied yet. Despite how it may appear, this application is not straightforward, as the unknown variables in the system are not only located in the numerator. In fact, the location of these variables also in the denominator makes the system of equations nonlinear and prevents the application of traditional observability techniques for this type of structure. To fill this gap, an algorithm for the observability analysis of 2D element structures was developed in this paper. The main contribution of this algorithm refers to the development of a new change in variables to linearize the system of equations and makes OM applicable for 2D element structures. The proposed methodology is characterized by its generality, as similar changes in variables can be used to linearize other nonlinear problems when the unknown variables appear in both the numerator and denominator of the system of equations.

To illustrate the applicability of the proposed algorithm, a state-of-the-art example (cantilever plate) was analyzed with the plane strain theory. This analysis included a step-by-step mathematical review of the system of equations, as well as a statistical study of the structural system identification with different measurement sets. The obtained results show that the unknown parameters (such as the Young's Modulus, E, and Poisson's Ratio, v) are successfully calculated with the proposed methodology.

Author Contributions: Conceptualization, H.M., J.A.L.G. and J.T.; methodology, B.M. and J.A.L.G.; software, B.M.; validation, H.M., J.A.L.G. and J.T.; formal analysis, B.M.; investigation, B.M. and J.A.L.G.; resources, H.M., J.A.L.G. and J.T.; data curation, J.A.L.G. and J.T.; writing—original draft preparation, B.M. and J.A.L.G.; writing—review and editing, H.M., J.A.L.G. and J.T.; visualization, B.M.; supervision, J.A.L.G. and J.T.; project administration, J.A.L.G.; funding acquisition, J.A.L.G. and J.T. All authors have read and agreed to the published version of the manuscript.

Funding: This research was funded by the Spanish Ministry of Economy and Competitiveness (grant number BIA2013-47290-R, BIA2017-86811-C2-1-R, and BIA2017-86811-C2-2-R), the Universidad de Castilla La Mancha (grant number 2018-COB-9092), and the Secretaria d' Universitats i Recerca de la Generalitat de Catalunya (grant number 2017 SGR 1481).

Institutional Review Board Statement: Not applicable.

Informed Consent Statement: Not applicable.

Data Availability Statement: All data during the study appear in the submitted article.

Acknowledgments: The authors are indebted to the Spanish Ministry of Economy and Competitiveness for the funding provided through the research projects BIA2013-47290-R, BIA2017-86811-C2-1-R, and BIA2017-86811-C2-2-R founded with FEDER funds. Funding for this research was provided to Behnam Mobaraki by the Spanish Ministry of Economy and Competitiveness through its program for his PhD. Part of this work was done through grant number 2018-COB-9092 from the Universidad de Castilla La Mancha (UCLM). The authors are also indebted to the Secretaria d' Universitats i Recerca de la Generalitat de Catalunya for the funding provided through Agaur (2017 SGR 1481).

Conflicts of Interest: The authors declare no conflict of interest.

Abbreviations

1D	One dimensional
2D	Two dimensional
[B]	Strain-displacement matrix
[C]	Coefficient matrix
CST	Constant Strain Triangle
[D]	Elastic matrix
E_j	Young's Modulus
E_N	Number of elements
$\{f\}$	Force vector
f_0	Subset of unknown forces
f_1	Subset of known forces
h	Plate thickness
H_i	Horizontal force at the ith node
I_j	Moment of inertia
[K]	Stiffness matrix
[K*]	Modified stiffness matrix
K_{ab}^*	Subset of the modified stiffness matrix
L_j	Length of the jth element
LST	Linear strain triangle
{N}	Vector of known parameters
N_N	Number of nodes in the FEM
NU_j^N	Change of variable for the negative sign in denominator
NU_j^p	Change of variable for the positive sign in denominator
OM	Observability method
u_i	Horizontal deflection at the ith node
[V]	Null space of the system of equations
V_i	Vertical force at the ith node
v_i	Vertical deflection at the ith node
$\{z\}$	General solution of the system of equations
ϑ	Poisson's Ratio
$\{\delta\}$	Vector of displacements
$\{\delta^*\}$	Modified vector of displacements
δ_0^*	Subset of unknown deflections
δ_1^*	Subset of known deflections
$\{\rho\}$	Vector of arbitrary values

References

1. Ceravolo, R.; Faraci, A.; Miraglia, G. Bayesian Calibration of Hysteretic Parameters with Consideration of the Model Discrepancy for Use in Seismic Structural Health Monitoring. *Appl. Sci.* **2020**, *10*, 5813. [CrossRef]
2. Park, J. Special feature vibration-based structural health monitoring. *Appl. Sci.* **2020**, *10*, 5139. [CrossRef]
3. Sirca, G.F.; Adeli, H. System identification in structural engineering. *Sci. Iran.* **2012**, *19*, 1355–1364. [CrossRef]

4. Liu, H.; Wang, X.; Tan, G.; He, X.; Luo, G. System reliability evaluation of prefabricated RC hollow slab bridges considering hinge joint damage based on modified AHP. *Appl. Sci.* **2019**, *9*, 4841. [CrossRef]
5. Campos, J.; Sharma, P.; Albano, M.; Ferreira, L.; Larrañaga, M. An Open Source Framework Approach to Support Condition Monitoring and Maintenance. *Appl. Sci.* **2020**, *10*, 6360. [CrossRef]
6. Romero, F.; Lofrano, E.; Paolone, A. Damage identification in parabolic arc via orthogonal empirical mode decomposition. In Proceedings of the International Design Engineering Technical Conference and Computers and Information in Engineering, Buffalo, NY, USA, 17–20 August 2014. [CrossRef]
7. Lofrano, E.; Romero, M.; Paolone, A. A pseudo-model structural damage index based on orthogonal empirical mode decomposition. *J. Mech. Sci.* **2019**. [CrossRef]
8. Wu, Z.; Liu, G.; Zhang, Z. Experimental study of structural damage identification based on modal parameters and decay ratio of acceleration signals. *Front. Archit. Civ. Eng. China* **2011**, *5*, 112–120. [CrossRef]
9. Salavati, M. Approximation of structural damping and input excitation force. *Front. Struct. Civ. Eng.* **2017**, *11*, 244–254. [CrossRef]
10. Xia, Y.; Lei, X.; Wang, P.; Liu, G.; Sun, L. Long-term performance monitoring and assessment of concrete beam bridges using neutral axis indicator. *Struct. Control Health Monit.* **2020**, *27*, e2637. [CrossRef]
11. Zhao, C.; Lavasan, A.A.; Barciaga, T.; Zarev, V.; Datcheva, M.; Schanz, T. Model validation and calibration via back analysis for mechanized tunnel simulations—The Western Scheldt tunnel case. *Comput. Geotech.* **2015**, *69*, 601–614. [CrossRef]
12. Xia, Y.; Wang, P.; Sun, L. Neutral Axis Position Based Health Monitoring and Condition Assessment Techniques for Concrete Box Girder Bridges. *Int. J. Struct. Stab. Dyn.* **2019**, *19*, 1940015. [CrossRef]
13. Yassine, R.; Salman, F.; Shaer, A.; Hammad, M.; Duhamel, D. Application of the recursive finite element approach on 2D periodic structures under harmonic vibration. *Computation* **2017**, *5*, 1. [CrossRef]
14. Hashash, Y.M.A.; Levasseur, S.; Osouli, A.; Finno, R.; Malecot, Y. Comparison of two inverse analysis techniques for learning deep excavation response. *Comput. Geotech.* **2010**, *37*, 323–333. [CrossRef]
15. Dudley, D.G.; Pao, H.Y. System identification for wireless propagation channels in tunnels. *IEEE Trans. Antennas Propag.* **2005**, *53*, 2400–2405. [CrossRef]
16. Sakurai, S.; Akutagawa, S.; Takeuchi, K.; Shinji, M.; Shimizu, N. Back analysis for tunnel engineering as a modern observational method. *Tunn. Undergr. Sp. Technol.* **2003**, *18*, 185–196. [CrossRef]
17. Sakurai, S.; Takeuchi, K. Back analysis of measured displacements of tunnels. *Rock Mech. Rock Eng.* **1983**, *16*, 173–180. [CrossRef]
18. Bhalla, S.; Yang, Y.W.; Zhao, J.; Soh, C.K. Structural health monitoring of underground facilities—Technological issues and challenges. *Tunn. Undergr. Sp. Technol.* **2005**, *20*, 487–500. [CrossRef]
19. Mobaraki, B.; Vaghefi, M. Effect of the soil type on the dynamic response of a tunnel under surface detonation. *Combust. Explos. Shock Waves* **2016**, *52*, 363–370. [CrossRef]
20. Mobaraki, B.; Vaghefi, M. Numerical study of the depth and cross-sectional shape of tunnel under surface explosion. *Tunn. Undergr. Sp. Technol.* **2015**, *47*, 114–122. [CrossRef]
21. Khamesi, H.; Torabi, S.R.; Mirzaei-Nasirabad, H.; Ghadiri, Z. Improving the Performance of Intelligent Back Analysis for Tunneling Using Optimized Fuzzy Systems: Case Study of the Karaj Subway Line 2 in Iran. *J. Comput. Civ. Eng.* **2014**, *29*, 05014010. [CrossRef]
22. Dehghan, A.N.; Shafiee, S.M.; Rezaei, F. 3-D stability analysis and design of the primary support of Karaj metro Tunnel: Based on convergence data and back analysis algorithm. *Eng. Geol.* **2012**, *141*, 141–149. [CrossRef]
23. Germoso, C.; Quaranta, G.; Duval, J.L.; Chinesta, F. Non-instrusive in-plane-out-of-plane seperated representation in 3D parametric elastodynamics. *Computation* **2020**, *8*, 78. [CrossRef]
24. Xia, Y.; Jian, X.; Yan, B.; Su, D. Infrastructure Safety Oriented Traffic Load Monitoring Using Multi-Sensor and Single Camera for Short and Medium Span Bridges. *Remote Sens.* **2019**, *11*, 2651. [CrossRef]
25. Vardakos, S.; Gutierrez, M.; Xia, C. Parameter identification in numerical modeling of tunneling using the Differential Evolution Genetic Algorithm (DEGA). *Tunn. Undergr. Sp. Technol.* **2012**, *28*, 109–123. [CrossRef]
26. Xiang, Z.; Swoboda, G.; Cen, Z. Optimal Layout of Displacement Measurements for Parameter Identification Process in Geomechanics. *Int. J. Geomech.* **2003**, *3*, 205–216. [CrossRef]
27. Santos, C.; Ledesma, A.; Gens, A. Backanalysis of measured movements in ageing tunnels. In *Geotechnical Aspects of Underground Construction in Soft Ground*; CRC Press: Boca Raton, FL, USA, 2012; pp. 223–229.
28. Lozano-Galant, J.A.; Nogal, M.; Castillo, E.; Turmo, J. Application of observability techniques to structural system identification. *Comput. Civ. Infrastruct. Eng.* **2013**, *28*, 434–450. [CrossRef]
29. Lozano-Galant, J.A.; Nogal, M.; Paya-Zaforteza, I.; Turmo, J. Structural system identification of cable-stayed bridges with observability techniques. *Struct. Infrastruct. Eng.* **2014**, *10*, 1331–1344. [CrossRef]
30. Castillo, E.; Lozano-Galant, J.A.; Nogal, M.; Turmo, J. New tool to help decision making in civil engineering. *J. Civ. Eng. Manag.* **2015**, *21*, 689–697. [CrossRef]
31. Nogal, M.; Lozano-Galant, J.A.; Turmo, J.; Castillo, E. Numerical damage identification of structures by observability techniques based on static loading tests. *Struct. Infrastruct. Eng.* **2016**, *12*, 1216–1227. [CrossRef]
32. Lei, J.; Nogal, M.; Lozano-Galant, J.A.; Xu, D.; Turmo, J. Constrained observability method in static structural system identification. *Struct. Control Health Monit.* **2018**, *25*, 1–15. [CrossRef]

33. Lei, J.; Xu, D.; Turmo, J. Static structural system identification for beam-like structures using compatibility conditions. *Struct. Control Health Monit.* **2017**, *25*, 1–15. [CrossRef]
34. Lozano-Galant, J.A.; Nogal, M.; Turmo, J.; Castillo, E. Selection of measurement sets in static structural identification of bridges using observability trees. *Comput. Concr.* **2015**, *15*, 771–794. [CrossRef]
35. Tomàs, D.; Lozano-Galant, J.A.; Ramos, G.; Turmo, J. Structural system identification of thin web bridges by observability techniques considering shear deformation. *Thin-Walled Struct.* **2018**, *123*, 282–293. [CrossRef]
36. Emadi, S.; Lozano-Galant, J.A.; Xia, Y.; Ramos, G.; Turmo, J. Structural system identification including shear deformation of composite bridges from vertical deflection. *Steel Compos. Struct.* **2019**, *32*, 731–741. [CrossRef]
37. Peng, T.; Casas, J.R.; Lozano-Galant, J.A.; Turmo, T. Constrained observability techniques for structural system identification using modal analysis. *J. Sound Vib.* **2020**, *479*. [CrossRef]
38. Rao, S.S. Analysis of plates. In *The Finite Element Method in Engineering*, 6th ed.; Elsevier: Amsterdam, The Netherlands, 2018; pp. 379–425. [CrossRef]
39. Castillo, E.; Nogal, M.; Lozano-Galant, J.A.; Turmo, J. Solving Some Special Cases of Monomial Ratio Equations Appearing Frequently in Physical and Engineering Problems. *Math. Probl. Eng.* **2016**, *2016*, 1–29. [CrossRef]
40. MathWorks. *MATLAB*; The MathWorks, Inc.: Natick, MA, USA, 2018.
41. Ramesh, S.S.; Wang, C.M.; Reddy, J.N.; Ang, K.K. Computation of stress resultants in plate bending problems using higher-order triangular elements. *Eng. Struct.* **2008**, *30*, 2687–2706. [CrossRef]
42. Yang, Y.; Tang, X.; Zheng, H. A three-node triangular element with continuous nodal stress. *Comput. Struct.* **2014**, *141*, 46–58. [CrossRef]
43. Piltner, R.; Taylor, R.L. Triangular finite elements with rotational degrees of freedom and enhanced strain modes. *Comput. Struct.* **2000**, *75*, 361–368. [CrossRef]
44. Neto, M.A.; Leal, R.P.; Yu, W. A triangular finite element with drilling degrees of freedom for static and dynamic analysis of smart laminated structures. *Comput. Struct.* **2012**, *108*, 61–74. [CrossRef]
45. Kattan, P.I. *MATLAB Guide to Finite Elements*; Springer: Berlin/Heidelberg, Germany, 2014. [CrossRef]

Article

Intelligent Discrimination Method Based on Digital Twins for Analyzing Sensitivity of Mechanical Parameters of Prestressed Cables

Zhansheng Liu [1,2,*], Guoliang Shi [1,2], Antong Jiang [1,2] and Wenjie Li [1,2]

[1] Faculty of Architecture, Civil and Transportation Engineering, Beijing University of Technology, Beijing 100124, China; shiguoliang@emails.bjut.edu.cn (G.S.); jiangat@emails.bjut.edu.cn (A.J.); liwj0827@emails.bjut.edu.cn (W.L.)

[2] The Key Laboratory of Urban Security and Disaster Engineering of the Ministry of Education, Beijing University of Technology, Beijing 100124, China

* Correspondence: lzs4216@163.com

Abstract: The information collected on large-span prestressed cables by field sensors is susceptible to interference, which leads to inaccurate collection of structural and mechanical parameters of large-span prestressed cables, resulting in misjudgment of structural safety performance. This paper proposes an intelligent judgment method for improving the sensitivity of analyzing mechanical parameters of prestressed cables based on digital twins (DTs). The safety performance of the structure was evaluated by analyzing the mechanical parameters. First, the information during prestressed cable tensioning is dynamically sensed, thereby establishing a multidimensional model of structural analysis. The virtual model is processed by the model modification rule to improve the robustness of the simulation; thus, a DT framework for the sensitivity judgment of the mechanical parameters of the cable is built. In the twin model, the simulation data of the real structure were extracted. Probabilistic analysis was performed using the Dempster–Shafer(D–S) evidence theory to discriminate the sensitivity of mechanical parameters of each cable node under the action of external forces with high accuracy and intelligence. Sensitivity analysis provides a reliable basis for the safety performance assessment of structures. Taking the wheel–spoke-type cable truss as an example, the application of DTs and D–S evidence theory to the sensitivity determination of cable mechanical parameters under temperature fully verified that the proposed intelligent method can effectively evaluate the safety performance of the actual structure.

Keywords: digital twin; Dempster–Shafer evidence theory; prestressed cable; sensitivity of mechanical parameter; intelligent discrimination; structural safety assessment

Citation: Liu, Z.; Shi, G.; Jiang, A.; Li, W. Intelligent Discrimination Method Based on Digital Twins for Analyzing Sensitivity of Mechanical Parameters of Prestressed Cables. *Appl. Sci.* **2021**, *11*, 1485. https://doi.org/10.3390/app11041485

Academic Editors: Hugo Filipe Pinheiro Rodrigues and Ivan Duvnjak

Received: 22 December 2020

Accepted: 3 February 2021

Published: 6 February 2021

Publisher's Note: MDPI stays neutral with regard to jurisdictional claims in published maps and institutional affiliations.

Copyright: © 2021 by the authors. Licensee MDPI, Basel, Switzerland. This article is an open access article distributed under the terms and conditions of the Creative Commons Attribution (CC BY) license (https://creativecommons.org/licenses/by/4.0/).

1. Introduction

With improvements in construction technology, the use of prestressed steel structures is widespread. With superior strength, flexible shape, and being light weight, prestressed space structures are commonly used in large public buildings [1]. In recent years, they have mostly been used in sports stadiums in China. The quality of construction of large-span spatial structures is also an important measure of a country's construction technology and capability, and the mechanical properties of the cables directly determine the overall performance of the structure [2]. Because large-span spatial structures are mostly used in buildings of high importance and large quantities are involved in the construction process, the safety requirements of the construction process are strict. Many experts and scholars have conducted a lot of research on the safety control of prestressed steel structures in the tensioning process.

Chen et al. [3] conducted research on a new type of tensioning construction forming method, construction error influence, and control technology for cable domes called the divisional lifting overall tensioning method. The application of this method during construction avoids large member displacement, thus improving the safety of construction. Guo et al. [4] investigated the effect of the initial cable length error in the prestressing state on the sensitivity of prestressed cables to length error. By controlling the length error, the prestress level during cable tensioning was effectively improved. Zhang et al. [5] proposed a joint-square double-brace structure to improve the force performance of the spatial structure and derived a formula for calculating the prestressing of the members of the structure under their self-weight, which provided theoretical support for the stability verification of the structure during the construction process. Xue et al. [6] studied a single-story saddle-shaped cross-cable network without an inner ring. By studying the nonlinear analysis of the forces of the structure based on the raw-dead unit method, the structural resistance to the continuous collapse of the inner-loop cable network was effectively improved. Liu et al. [7] used the response surface method and Monte Carlo method to calculate the structural reliability index and obtain the effect of tensioning cable relaxation at different locations on structural reliability, which allows the calculation of prestress loss in spatial structures and provides a basis for the safety assessment of structures. Asadolahi, SM. et al. [8] proposed equations to estimate the joint stiffness and energy dissipation reduction of the system by analyzing it for two time periods, during and after the application of prestress. Arezki et al. [9] investigated the effect of temperature variation on the safety performance of cable truss structures and cables. Castillo et al. [10] derived the damage accumulation process of a cable system under normal loading and established a theoretical model for the fatigue life of a cable considering load redistribution. Shekastehband et al. [11] conducted a theoretical analysis and experimental study on the sudden breakage of cables in tensioned monolithic structures, pointing out that severe cases can cause a continuous collapse of the tensioned monolithic structure.

There are several shortcomings in the component sensitivity discrimination and safety performance assessment of prestressed steel structures in the above studies. (i) The safety assessment of structures relies on data collected by field sensors, resulting in insufficient accuracy of structural performance analysis. (ii) The damage level of the structure cannot be displayed visually, resulting in insufficient information for structural performance analysis. With the development of the new generation of information technology and the promotion of industrial information systems, the application of technologies such as Digital twins (DTs) to engineering construction has become an increasingly researched topic [12]. Applying DTs to engineering practice can significantly improve the accuracy and intelligence of structural performance analysis.

DT technology is a technique that simulates the state and behavior of physical entities portrayed in a dynamic virtual model with high fidelity [13]. Zhang et al. [14] combined DTs and virtual reality technology to propose an intelligent control method for underground equipment, which solved the problem of difficulty in remotely controlling cantilevered road heading machines. Gong et al. [15] proposed an intelligent regulation and control system for air flow at the outlet of the digging face by analyzing the DT technology, thereby realizing the mapping interaction between the physical entity and the virtual twin in the coal mining process. Meng et al. [16] used DT technology in vehicle life prediction and maintenance decision making, which describes the working state of the system more accurately, thus enabling online optimal decision making and feedback control of the vehicle. Szpytko et al. [17] proposed an integrated maintenance decision model for cranes under operating conditions based on the DT concept, which effectively improved the operational efficiency of cranes and reduced the safety risk of equipment operation. Zhou et al. [18] proposed a DT model of a centrifugal impeller and established parametric centrifugal impeller DTs, which can significantly improve the design and manufacturing level of the centrifugal impeller and shorten the design cycle. Yu et al. [19] proposed a DT framework for solving the health state monitoring problem of complex systems, which improves the accuracy and reduces the uncertainty of analysis for the health monitoring of complex systems in intelligent manufacturing. Krishnan et al. [20] used the DT concept for the detection of electric drive faults, which improved the performance analysis accuracy of electric vehicles during usage. Seon et al. [21] conducted a study to accurately analyze the structural integrity and remaining service life of aircraft. Based on the study of high-fidelity nondestructive inspection techniques, a DT of the structure was created to optimize the monitoring mechanism, which provided a reference for the performance analysis of complex structures. This shows that DT technology has been studied and widely applied in manufacturing.

Owing to the lack of research on the assessment of structural safety performance during tensioning of prestressed steel structures and combined with the status of research on DTs, this paper proposes an intelligent discrimination method based on DTs for the sensitivity analysis of mechanical parameters of prestressed cables. The safety performance of the structure is intelligently evaluated through the analysis of the sensitivity of the component mechanical parameters. First, based on the concept of DTs, a framework for sensitivity discrimination was established, and the real tension was dynamically perceived and simulated. In the virtual model, the mechanical parameters of the structure can be extracted directly, and the mechanical properties of the structure can be displayed intelligently. After extracting the mechanical parameters, the Dempster–Shafer (D–S) evidence theory was fused to process the structural mechanical performance information. Therefore, the change probability of the mechanical parameters of each node of the cable was accurately analyzed, and the sensitivity of the prestressed cable was intelligently determined, thus providing a basis for the safety performance evaluation of the structure. Driven by DTs, the D–S evidence theory is integrated to realize intelligent discrimination of the sensitivity of the mechanical parameters of the prestressed cable. Based on the above theoretical method, application was made to a wheel–spoke cable truss as an example. In this study, the sensitivity of the mechanical parameters of the cable under the effect of temperature during tensioning was analyzed, and the effectiveness of this method was verified.

2. Construction of a DTs Framework for Sensitivity Discrimination

Prestressed steel structures have the highest safety risk during the construction phase because of the incompleteness of the structure, time-varying nature of the material, complexity of the loads applied, and immaturity of the structural resistance. Mechanical analysis during cable tensioning is in the domain of construction mechanics, where the geometric, physical, and boundary parameters are functions of time. The safety assessment of the tensioning process is a multidimensional mechanical problem of the coupling time and space. Combined with the characteristics of tensioning safety assessment, this study considers the sensitivity of the mechanical parameters of tension cables as an entry point. The reliability of the structure is determined by analyzing the sensitivity of the mechanical parameters at each point in the member. Supported by the concept of DTs, the dynamic perception and simulation of realistic tension are performed, and the mechanical performance state of the structure is intelligently displayed by the twin model.

2.1. Capture of Twin Information

The DTs replicate a real physical entity through visual virtual space modeling, and the virtual model simulates the dynamic behavior of that entity in a real environment [22,23]. It is necessary to compare multidimensional structural data during assessment of the structural safety state [24]. Therefore, the first step in the twin modeling of cable tensioning is to identify the factors influencing the safety for analysis. The changes in the external effect and structural mechanical properties of the prestressed steel structure subjected to the tensioning process are the external and internal causes of the structural state change. The changes in the structure and mechanical performance of the structure can directly change the state of the structure. Likewise, the changes in the structural state of the prestressed steel structure in the tensioning process will also cause a change in the structural mechanical properties; thus, the changes in the structural state can be reflected in the degree of changes in the structural mechanical properties. The state of the structure directly responds to the safety performance of the structure. The intrinsic relationship between the basic factors influencing the safety performance of prestressed steel structures during tensioning is shown in Figure 1.

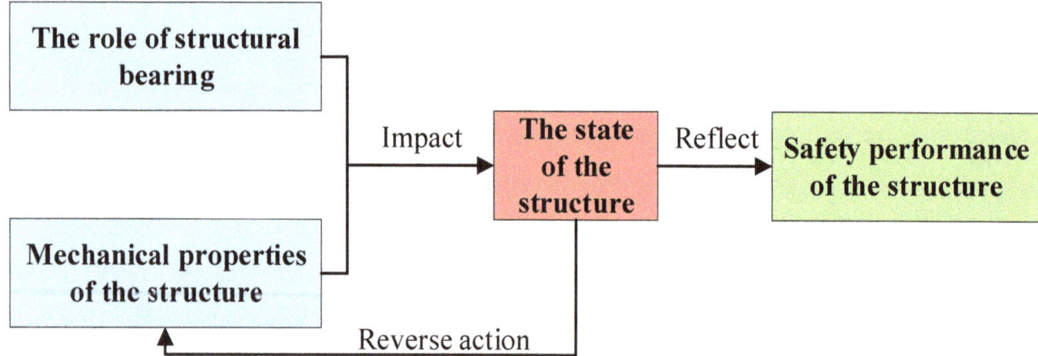

Figure 1. Intrinsic relationship of basic influencing factors of safety performance of prestressed steel structures during tensioning.

From the intrinsic relationship of the factors influencing the structural safety performance, it can be concluded that the information to be captured for the tensioning process is mainly divided into the action to which the structure is subjected (S) and the mechanical properties of the structure (R). The twin model established in this study can accurately map the state of the structure.

1.   The role of structure bearing

There are multiple causes of structural damage or destruction during the tensioning process. By combining the effects of various external influences [25], this study focuses on capturing and sensing the member length errors (L_e), wind load effects (W_l), and temperature effects (T_e). The component length error is the most important factor causing the initial form and pretension deviation of the cable tension structure [26]. The effective control of the length error effect is key in the design of prestressed steel structures, especially in the design of construction tensioning. In an actual project, the geometric error of the cable tension structure mainly includes the error of the release length of the cable, error of the size of the anchorage or node, and deviation of the bearing position. Geometric errors can cause deviations in the initial form of the flexible pretensioned structure design, resulting in inconsistencies between construction and design and a consequent loss in prestress. The severity and sensitivity of the wind vibration effects of large-span space structures are no less than those of high-rise and towering structures. In most cases, wind loads tend to be controlling loads in the design of structures [27]. The cable is a flexible system, which very easily deforms under the wind load, thus affecting the safety performance of the entire structure. Therefore, the monitoring of wind pressure helps to scientifically predict the safety performance of cables. Because steel structures have temperature-sensitive properties, the analysis of temperature effects is very important in the design and construction of large-span steel structures [28]. Changes in temperature result in temperature stresses and temperature deformations in the structure. Prestressed cables are important elements of superstationary structures, and the effect of temperature inevitably causes changes in the internal forces of the cables, thus affecting the safety performance of the structure.

2.   Mechanical properties of the structure

It is more common to determine the safety performance of a structure according to the state of the prestressed steel structure during tensioning. The state of the structure refers to all the forms of the structure and its materials, and the state of the structure is influenced by the mechanical properties of the structure. Based on the mechanical properties of the cables, the sensitivity of the mechanical parameters at each point can be discerned, and the reliability of the structure can be assessed. The key elements in the analysis of prestressed steel structures during tensioning are the mechanical property indices, such as cable force (C_f), deflection (ω), stress (δ), strain (ε), and crack (C), which are closely related to the reliability of the construction process of the structure.

2.2. Multidimensional Modeling for Intelligent Discrimination of Sensitivity to Mechanical Parameters of Cables

In this study, the structural information of cable tensioning is derived from the finite element model of the structure, which is highly integrated with each construction element of the prestressed steel tensioning process and contains sufficient twin data to support the analysis of the sensitivity of the member mechanical parameters. This allows an intelligent analysis of the safety performance of structures. Moreover, finite element analysis has been widely used as a better mechanical simulation tool in the structural construction industry [29]. The twin information is divided into two categories based on the analysis of the information that needs to be captured for tension safety assessment: the action to which the structure is subjected and the mechanical properties of the structure. These are expressed by Equation (1) and Equation (2). As important simulation information for the twin model, the action on the structure (S), and the mechanical properties of the structure (R) are expressed specifically as:

$$S = (L_e, W_l, T_e) \tag{1}$$

$$R = (C_f, \omega, \delta, \varepsilon, C) \tag{2}$$

In Equation (1), L_e denotes the member length errors, W_l denotes the wind load effects, and T_e denotes the temperature effects. In Equation (2), C_f denotes the cable force, ω denotes deflection, δ denotes stress, ε denotes strain, and C denotes a crack.

During the creation of the twin model, the action to which the structure is subjected and the mechanical properties of the structure are collected in real time using sensors. At the same time, a virtual model of the real structure is created during the finite element analysis. In the virtual model, the twin data of the effect of the structure and the mechanical properties of the structure are simulated by setting the corresponding working conditions. This enables the twin simulation of the realized tensioning process and establishes a model basis for the intelligent discrimination of the sensitivity of the mechanical parameters of the prestressed cables driven by the DTs. This in turn provides a basis for the analysis of the safety performance of structure construction. A multidimensional model for the intelligent discrimination of the sensitivity of the mechanical parameters of the cables is represented by Equation (3). The multidimensional model consists of five dimensions, expressed as follows:

$$DTM = (S_{pr}, S_{vm}, L_{dp}, L_{fa}, C_n) \tag{3}$$

In Equation (3), DTM denotes a multidimensional model for the intelligent discrimination of the sensitivity of the mechanical parameters of the cables, S_{pr} denotes the physical structure entity, S_{vm} denotes the virtual structure model, L_{dp} denotes twin data processing layer, L_{fa} denotes the functional application layer, and C_n denotes the connection between the components. The multidimensional twin model enables the simulation and mapping of realistic tensioning processes. By processing the structural parameters from realistic monitoring and virtual simulation, the sensitivity of the mechanical parameters of the structural components can be determined. As a result, the safety performance of the structure can be assessed, and precise maintenance of the construction site can be achieved. The multidimensional model for the intelligent discrimination of the sensitivity of the mechanical parameters of the oriented cables is shown in Figure 2.

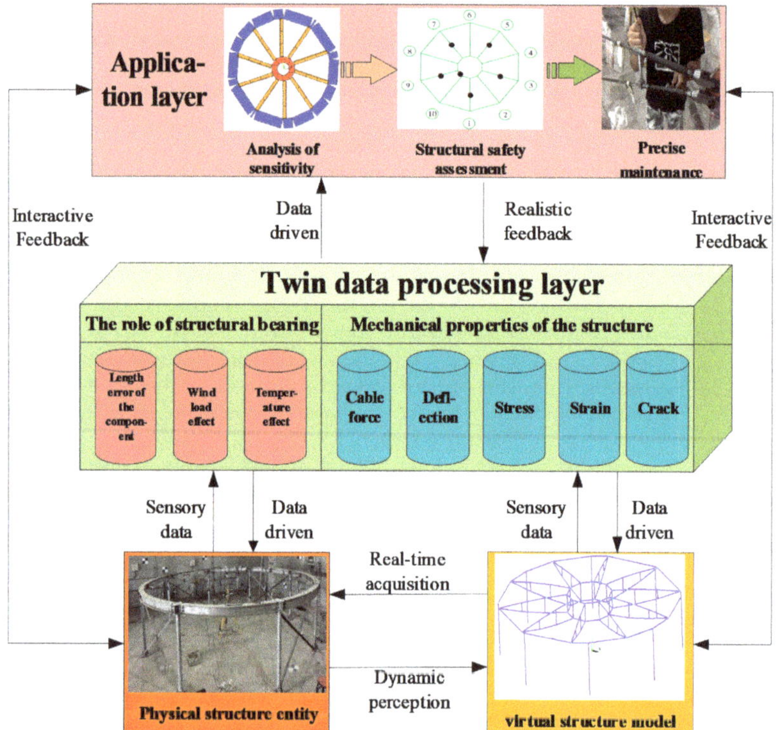

Figure 2. Multidimensional model for intelligent discrimination of sensitivity to mechanical parameters of cables.

2.3. Correction of the Virtual Model

Under the condition of establishing a multidimensional model, to enhance the simulation capability of the virtual model, it is necessary to overcome the high fidelity of the model and correct the virtual twin model of the tensioning process of the cables. The average weighting method [30] was used to fuse the monitored data of realistic tensioning and the simulated data of the finite element model to correct the virtual model. The basic principle involves selecting different weights for different location sensitivity indicators to achieve the goal of minimizing the sum of squared Euclidean distances between the fusion results and each sensitivity indicator.

Suppose that the actual monitored values of a mechanical parameter at each node of a cable in a certain working condition are x_1, x_2, ..., x_n. Then, the average value $\bar{x}$ of the actual monitored values is expressed as:

$$\bar{x} = \sum_{i=1}^{n} x_i / n \tag{4}$$

Then the weight ω_i of each mechanical parameter is represented by Equation (5), which is expressed as:

$$\omega_i = \left(\frac{1}{d_i}\right) / \sum_{i=1}^{n}\left(\frac{1}{d_i}\right) \tag{5}$$

In Equation (5), d_i is the Euclidean distance from each mechanical parameter to the mean of the mechanical parameters, expressed as:

$$d_i = \|\bar{x} - x_i\| \tag{6}$$

From this, the weighted average of the monitored values of mechanical parameters ($\hat{x}$) for the entire cable can be calculated by combining the data of each node, that is,

$$\hat{x} = \sum_{i=1}^{n} \omega_i x_i \tag{7}$$

The overall monitoring value D of the structure can be calculated from the weighted average of the monitoring values of the mechanical parameters of each cable, that is:

$$D = \mu + \alpha_c \sigma \tag{8}$$

α_c is the confidence level of the mechanical parameter analysis, 1.5 in this study. μ and σ denote the mean and standard deviation of the weighted average of the monitored values of the mechanical parameters ($\hat{x}$) for each cable, respectively.

Similarly, using the above steps, the simulation values D^* of the mechanical parameters in the finite element model can be calculated. Using Equation (9) to judge the fidelity of the simulation model, the correction of the virtual model is completed, which ensures that the simulation data effectively represent the mechanical properties of the real structure:

$$E_D = \frac{|D^* - D|}{D} \times 100\% \tag{9}$$

In Equation (9), E_D is a metric for determining the fidelity of the twin model.

2.4. DTs-Driven Sensitivity Intelligence Discriminative Framework

By establishing a high-fidelity virtual model, it is possible to simulate the mechanical properties of cables in a virtual space for realistic tensioning processes. The simulation is carried out in the virtual model for the conditions in the real tensioning process to obtain highly consistent mechanical property information with the real structure. To improve the sensitivity analysis accuracy of the mechanical parameters of the cables, multiple types of simulation data were analyzed by integrating the D–S evidence theory. The sensitivity of the mechanical parameters of each node on the cable can be intelligently discerned to provide a basis for the safety performance assessment of the structure, which in turn ensures the feasibility of the maintenance of the real structure. The intelligent discriminative framework for the sensitivity of the mechanical parameters driven by DTs is shown in Figure 3.

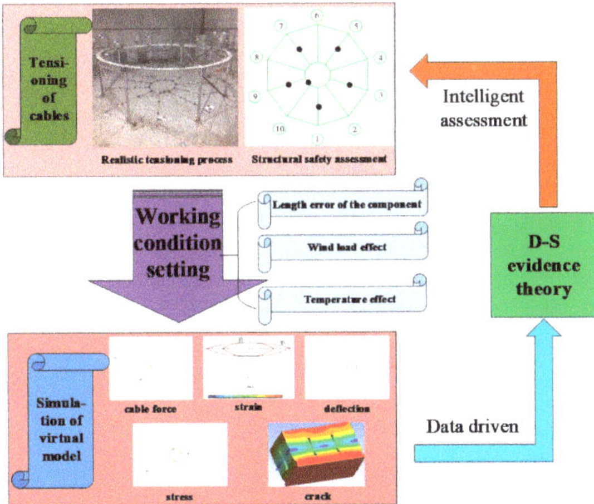

Figure 3. Intelligent discriminative framework for mechanical parameter sensitivity driven by digital twins.

3. Analysis of Simulation Data Based on D–S Evidence Theory

By establishing a high-fidelity twin model, the simulated data in the model can be directly extracted for analysis, which realizes intelligent discrimination of the sensitivity of the cable mechanical parameters based on the DTs. By integrating the D–S evidence theory to process the information about the structural mechanical properties, the probability of mechanical property change at each point of the cables is calculated to discern the sensitivity of mechanical parameters at each point and provide a basis for the safety performance assessment of the structure. D–S theory, as a technique for the fusion of multisource information, belongs to the category of artificial intelligence [31] and has been applied in business administration and road transportation. In this study, it was applied to discriminate the sensitivity of the mechanical parameters of a structure to improve the accuracy of the safety assessment.

3.1. D–S Evidence Matrix

The basic principle of the D–S evidence matrix is as follows: based on multiple sources of heterogeneous data, multiple mechanical parameter sensitivity discriminators are identified independently, and the basic probability assignment function is determined by the probability assignment of each mechanical parameter sensitivity discriminator. The final discriminatory results are obtained by fusion using D–S synthesis rules [32].

Assume that Ω is the discriminative framework for the sensitivity of the mechanical parameters of the prestressed cable members, which is expressed by Equation (10), and the specific expression is:

$$\Omega = (e_1, e_2, \cdots, e_n) \tag{10}$$

In Equation (10), $e_i (i = 1, 2, ..., n)$ expresses the change in the mechanical properties of the ith node of the prestressed cable.

All possible working conditions of the prestressed cable element are the power set "2^{Ω}", which is expressed as Equation (11):

$$2^{\Omega} = \left(\varnothing, e_1, e_2, \cdots, e_n, e_1 \cup e_2, e_i \cup e_j \cup \cdots \cup e_k \cdots\right) \tag{11}$$

In Equation (11), $\varnothing$ indicates no change in the mechanical properties of the prestressed cable, and e_i indicates a change in the mechanical properties of a single node of the prestressed cable. $e_i \cup e_j \cup \cdots \cup e_k$ indicates the change of mechanical properties of multiple nodes of the prestressed cable.

The basic probability distribution function (mass function) can be expressed as m: $2^{\Omega} \rightarrow [0,1]$, satisfying Equation (12):

$$\begin{cases} m(\varnothing) = 0 \\ \sum_{A \subset \Omega} m(A) = 1 \end{cases} \tag{12}$$

In Equation (12), A is a certain working condition of the prestressed cable, and $m(A)$ is the basic probability distribution function of A.

Assume that $m_j(A_i)$ is the basic probability distribution function for the jth mechanical parameter sensitivity indicator. To determine the change in the mechanical properties of the ith node, the synthesis rule of the D–S evidence matrix is Equation (13):

$$DF(A) = m_1 \oplus m_2 \oplus \cdots \oplus m_n(A) \begin{cases} \dfrac{\sum_{\cap A_i = A} \prod_{1 \leq j \leq n} m_j(A_i)}{K} (A \neq \varnothing) \\ 0 \ (A = \varnothing) \end{cases} \tag{13}$$

In Equation (13), $K = \sum_{A_1 \cap A_2 \cap \cdots \cap A_n \neq \varnothing} m_1(A)_1 m_2(A)_2 \cdots m_n(A)_n = 1 - \sum_{A_1 \cap A_2 \cap \cdots \cap A_n = \varnothing} m_1(A_1) m_2(A_2) \cdots m_n(A_n)$.

3.2. Analysis Process of Sensitivity

The DT framework is integrated with the D–S evidence theory to intelligently discriminate the sensitivity of the mechanical parameters of the cables. First, a twin model was built for a realistic tensioned structure, and the model was modified by the weighted average method to improve the fidelity of the model. In the virtual model with high fidelity, the working conditions corresponding to the construction site were set, and the mechanical performance index data simulated in the model were extracted. The relevant information of the index data is determined probabilistically, and the sensitivity of the integrated mechanical parameters of each node of the cable is calculated by integrating the D–S evidence theory algorithm. In this study, the sensitivity of the mechanical parameters is characterized by the degree of change in the mechanical parameters of each node on the cable. By sorting the probability of change in mechanical parameters at each point of the cable, the change degree of mechanical properties at each point can be determined. According to the sensitivity of the mechanical parameters, the most vulnerable position of the structure can be directly determined, thus providing a basis for the safety performance evaluation and maintenance of the structure. The process of safety performance assessment and maintenance of the structure through a sensitivity analysis of the mechanical parameters is shown in Figure 4.

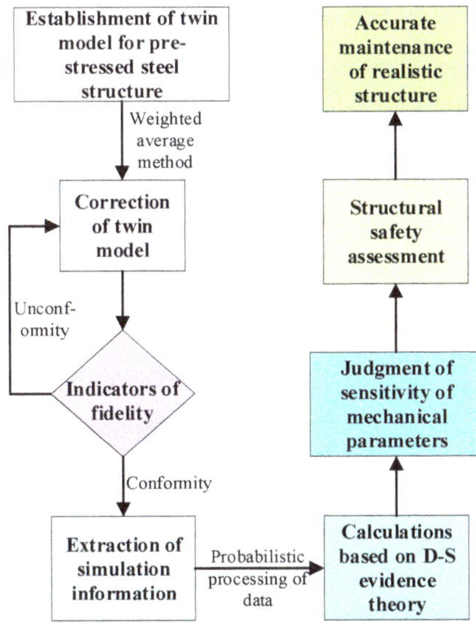

Figure 4. Safety performance assessment and maintenance process for structures.

4. Intelligent Discrimination of the Sensitivity of Mechanical Parameters of Prestressed Cables under the Effect of Temperature

In the intelligent discrimination framework of mechanical parameter sensitivity driven by DTs, a high-fidelity virtual model of a realistic tensioning structure is established. The twin data are also processed by integrating the D–S evidence theory, which can accurately discriminate the sensitivity of the mechanical parameters of each prestressed cable node under the effect of an external environment using model simulation data. The experimental model built in this study is a reduced-scale test model based on a certain wheel–spoke cable truss project. Compared with the actual project, the scale ratio of the test model was 1:10, cross-sectional area ratio of the cable was 1:100, and materials were identical. The structure span of the test model was 6 m and consisted of 10 radial cables, ring cables, braces, nodes, outer ring beams, and steel columns. The radial cables include upper and lower radial cableshe radial cables include upper and lower radial cables, and the ring cables include upper and lower ring cables.

The struts included the outer, middle, and inner struts. The model used to test the wheel, spoke cable truss, is shown in Figure 5. This study focuses on the analysis of changes in structural mechanical parameters under the effect of temperature difference. First, the mechanical parameters of the structure are simulated in the high-fidelity virtual model, and the degree of change in the mechanical parameters of each node is analyzed by fusing the D–S evidence theory to find and monitor the most sensitive position, thus realizing the intelligent evaluation of the safety performance of the structure. In the test process, to simulate roof installation after the cable is formed, the effect of temperature difference on the structural safety performance of the cable is considered, and the extreme working conditions are considered. Combined with the temperature change in the region where a practical project is located, it is concluded that the annual temperature change in the region is 55 °C. In order to simulate the state of the real structure, a test model is built. According to the constructed test model, the working condition was set to a temperature rise of 55 °C. The initial temperature value of the structure is given first, and then the final temperature value of the structure is set by adjusting the temperature so that the temperature change is 55 °C. At the same time, the temperature change is set to increase by 1 °C per minute, and the mechanical properties of the structure are analyzed when the temperature increases by 55 °C, thus achieving consistency with realistic structure state. Sensors are arranged in the field for real-time monitoring, whereas the virtual model combined with the D–S evidence theory discriminates the sensitivity of the mechanical parameters of the cables under the effect of temperature.

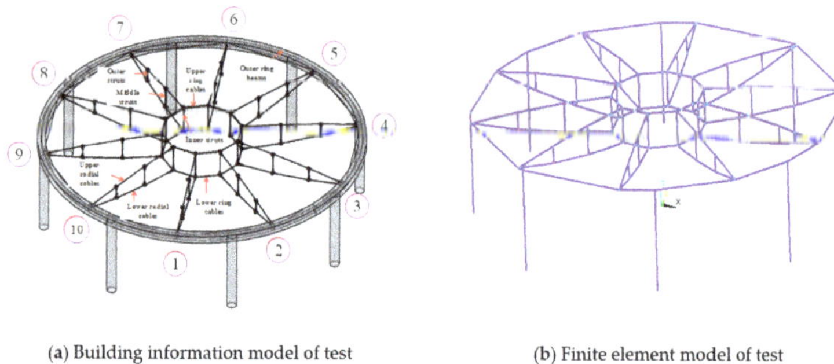

(a) Building information model of test   (b) Finite element model of test

Figure 5. Model for the test of the wheel-spoke cable truss.

4.1. The Analysis Process of Sensitivity

A high-fidelity twin model is the basis for the intelligent discrimination of the sensitivity of mechanical parameters in cables under the effect of temperature. In this study, ANSYS was used to establish a finite element model for the simulation and analysis of structural performance. In the finite element model, the mechanical performance indices such as cable force (C_f), deflection (ω), stress (δ), strain (ε), and crack (C) of the structure can be simulated by setting various types of effects to which the structure is subjected. Among them, the main effects of the structure include the length error of the member (L_e), wind load action (W_l), and temperature action (T_e).

After the finite element model is established, to achieve a realistic mapping of the real structure, the model needs to be modified to improve the fidelity of the model. The monitored data of the field sensors and the simulated data of the model were fused according to the model correction law proposed in Section 2.3. Finally, the parameters and node connections of the twin model were adjusted using the cable force as the control object. In the performance analysis of this structure, the analysis focused on the support and midspan position of each section of the cable. Under the self-weight of the structure, cable force sensors are arranged at the supports and midspan positions of each section of the upper and lower radial cables of the test structure to collect the cable forces of the cable members. The cable forces of the corresponding nodes of the cables are simulated in the twin model. In this case, the nodes of the real structure with the sensors are the same as the nodes of the simulated data in the model. Taking the upper radial cable of the first bay as an example, the nodes where the sensors are arranged in the real structure and the nodes simulated in the model are shown in Figure 6. A total of seven nodes were selected in upper radial cable 1. The cable force and strain sensors are arranged at each control node, while the real-time acquisition of vertical displacement is carried out by a total station. In the finite element model, the analysis of the corresponding mechanical parameters is carried out for each node. Using the average weighting method, the index (E_D) for adjusting the fidelity of the twin model is within 5%, and mapping to the real structure can be achieved.

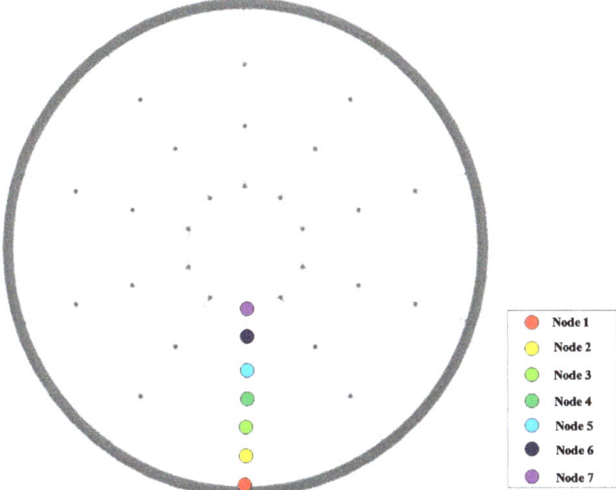

Figure 6. Schematic diagram of the selection of nodes.

In this study, the finite element model was modified by adjusting the cross-sectional area of the cable before studying the effect of temperature on the mechanical properties of the structure. In Figure 6, by collecting the cable force of each node under the action of self-weight, the weighted average method is used for data fusion. Considering that the change of the section size of the component will cause the change of the self-weight of the structure, the fidelity of the finite element model is improved by modifying the section area of the component. The dimensions of the modified finite element model members are listed in Table 1. Taking the whole of each cable as the object of examination, a comparison of the simulation of the cable force before and after the model correction is shown in Table 2.

Table 1. Dimensions of the modified finite element model members.

Member Type	Location	Cross-Sectional Area (mm^2)
Radial cable	Upper level	24.6
	Lower level	33.3
Ring cable	Upper circle	24.6
	Lower circle	49.1
Strut	Outer circle	62.8
	Mid circle	62.8
	Inner circle	62.8
Ring beam	Outer circle	4300

Table 2. Comparison of simulation degree of cable force before and after model correction.

Component Unit Number	Monitoring Values of the Structure (N)	Simulation Value before Correction (N)	Simulation Value after Correction (N)	Error before Correction	Error after Correction
Upper radial cable 1	4890	5572	4775	14%	−2%
Upper radial cable 3	5850	5563	5713	−5%	−2%
Upper radial cable 5	5150	5570	5087	8%	−1%
Upper radial cable 7	5660	5569	5772	−2%	2%
Upper radial cable 9	5140	5573	5137	8%	0%
Lower radial cable 1	4260	4481	4136	5%	−3%
Lower radial cable 3	4790	4473	4851	−7%	1%
Lower radial cable 5	4100	4569	4134	11%	1%
Lower radial cable 7	4490	4368	4655	−3%	2%
Lower radial cable 9	4210	4481	4302	6%	2%

From the comparison before and after the model correction, it can be seen that the revised model can reflect the state of the real structure effectively. The data of multiple nodes of each cable are fused based on the average weighting method in the correction process to ensure that the assurance index of the virtual model is within 5%. Therefore, the error of the corrected simulation is lower when each cable is studied as a whole. In this study, it can be concluded from Table 1 that the error between the simulated data of the modified model and the actual collected data is within 3%. The resulting finite element model can realistically and accurately map the performance changes in the structure.

4.2. Data Analysis Based on D–S Evidence Theory

Based on the establishment of a high fidelity finite element model, the data from the simulation of multiple mechanical parameters of the structure were fused. The sensitivity of the mechanical parameters at each node of the cable under the effect of temperature was analyzed using the probabilistic treatment and D–S evidence theory. In this study, to verify the accuracy of the judgment based on the D–S evidence theory, the points where the simulated data are collected in the model are set to be the same as the locations of the monitored points arranged in the real cables. Upper radial cable 1 is the object of study, and there are seven monitored and simulated points arranged at the support and midspan positions of each section of the cable. The selection of the monitored and simulated points is the same as that in Figure 6.

In the finite element model, a condition consistent with the real tensioning process was set, that is, a temperature rise of 55 °C. The simulated values of the cable force, strain, and deflection at each node were extracted from the model, and the sensitivity of the mechanical parameters at each point under the effect of temperature was calculated by probabilistic processing. In calculating the probability of the simulated values at each point, the analysis using the D–S evidence theory should ensure that the sum of the probabilities of parameter changes at each node is 1 when a single parameter is used as the discriminant.

For example, when using the cable force as the basis for discrimination, the probability of the change in the cable force at each node is calculated using Equation (14).

$$P_i^* = \frac{\left| C_{fi} - C_{fi}^* \right|}{C_{fi}} \tag{14}$$

In Equation (14), C_{fi} indicates the value of the cable force at each node on the cable before the effect of temperature, C_{fi}^* and indicates the value of the cable force at each node on the cable after a temperature increase of 55 °C. To ensure that the sum of the damage probability of each node on the cable is 1 when the cable force is used as the basis for discrimination, the probability calculated by Equation (14) needs to be corrected. The probability (P_i) of the change in the cable force at each node after correction was calculated according to Equation (15):

$$P_i = \frac{P_i^*}{\lambda} \tag{15}$$

In Equation (15), λ denotes the coefficient of the correction of probability $\lambda = \sum_{i=1}^{7} P_i^*$. The probabilities calculated accordingly can be used in the fusion analysis of the D–S evidence theory to discriminate the sensitivity of the mechanical parameters of each node.

Similarly, the probability of change in the mechanical parameters at each point when strain and deflection are used as discriminators can be derived. Combining the D–S evidence theory to fuse various types of data and synthesize the sensitivity of each node can lead to the response of the cable to the effect of temperature. At the same time, the rate of change of the cross sectional area of each point on the real structure is used as an important basis for the detected results of structural safety performance. The data collection of the rate of change of the cross-sectional area of each node of the cable in the field is shown in Figure 7.

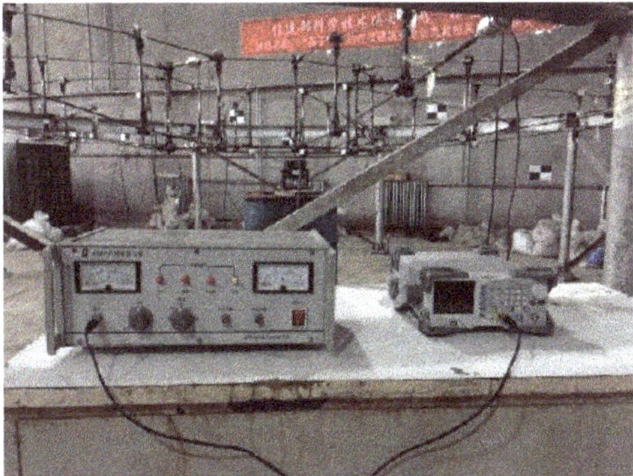

Figure 7. Data acquisition of the rate of change of cross-sectional area.

The rate of change of the cross-sectional area of each node on the cables was calculated according to Equation (16). To make the comparison of the results clearer, this study adjusted the sum of the detection result values to 1 for each point. The rate of change of the cross-sectional area of each node on the cable was corrected according to Equation (17), and the results of the detection of the structural safety performance expressed by the corrected rate of change of the cross-sectional area of each node on the cable were used for comparison with the results of the D–S evidence theory. The discrimination of the sensitivity of each node of the cable is presented in Table 3.

$$D_{mi}^* = \frac{\left| A_{mi} - A_{mi}^d \right|}{A_{mi}} \tag{16}$$

Table 3. Judgment of the sensitivity of each node of the cable.

Location	Discrimination Based on Cable Force	Discrimination Based on Strain	Discrimination Based on Deflection	Discriminative Results of D–S Evidence Theory	Detected Results of Structural Safety Performance
Node 1	12.5%	12.3%	11.2%	8.21%	8.69%
Node 2	13.1%	12.8%	12.3%	9.84%	9.24%
Node 3	13.7%	13.7%	12.7%	11.37%	10.72%
Node 4	15.6%	15.9%	14.9%	17.63%	18.67%
Node 5	16.4%	15.6%	15.6%	19.04%	20.18%
Node 6	14.9%	15.3%	16.9%	18.38%	19.43%
Node 7	13.8%	14.4%	16.4%	15.53%	14.68%

In Equation (16), D^*_{mi} denotes the rate of change of the cross-sectional area of each node of the cable in a realistic structure, and A_{mi} denotes the cross-sectional area of each node on the cable before the effect of temperature. A^d_{mi} denotes the cross-sectional area of the nodes on the cable after a temperature increase of 55 °C.

$$D_{mi} = \frac{D^*_{mi}}{k} \qquad (17)$$

In Equation (17), D_{mi} denotes the corrected rate of change of the cross-sectional area of each node on the cable, and μ indicates the coefficient of correction of the rate of change, $k = \sum_{i=1}^{7} D^*_{mi}$.

In Table 3, by comparing the change rate of mechanical parameters of each point based on D–S evidence theory analysis and the change rate of area based on field acquisition, it can be obtained that when the temperature increases by 55 °C, the order of sensitivity of each point is consistent, and the sensitivity of node 5 is the highest, so it is necessary to focus on monitoring the node.

4.3. Analysis of Experimental Results

In the modification of the finite element model, it is possible to effectively target a structural parameter as the control object; however, the other simulated parameters are not necessarily guaranteed to be consistent with the real structure. As the change in several mechanical parameters will have an impact on the cross-sectional area of the components and each mechanical parameter is also taken into consideration when probing the safety performance of the real structure, it is necessary to make a comprehensive judgment on the various types of data from the simulation by means of data fusion. The probability of mechanical property change at each node of the cable can be determined more precisely by the D–S evidence theory incorporating various sensitivity discriminant parameters. By comparing the detected results with the safety performance of the real structure, the ranking of the damage degree of each node is consistent, and the synchronization of the structural sensitivity degree of each point analyzed with the real monitoring value is above 95%, which provides a reliable basis for the safety assessment of the structure. In this study, with a temperature increase of 55 °C as the working condition, it was determined that the most sensitive position of the mechanical parameters of upper radial cable 1 under the effect of temperature is at node 5. This needs to be closely monitored to ensure the reliability of the structure.

To verify the validity of the proposed method, the fidelity of the model and the number of discriminative indicators were varied in this study, and experimental analyses were performed on other cables as well. During the test, the mechanical parameters of each cable were analyzed. In this paper, the upper radial cable 1 is described as the key analysis object. The average weighting method is used to modify the virtual model, whereas the D–S evidence theory is used to integrate various types of discriminant indices for the sensitivity analysis of structural mechanical parameters, which are finally applied to the analysis of the safety performance of the entire structure. The fidelity of the model and the number of discriminative indicators were adjusted in the sensitivity analysis of the individual cable mechanical parameters for the entire structure. By analyzing the influence of varying fidelity and number of discriminative indicators on the accuracy of the analysis results, the relationship between accuracy and the degree of assurance and number of discriminant indicators was obtained, as shown in Figure 8. From the figure, it can be concluded that the fidelity must be within 6%, and the number of discriminative indicators must not be less than 3 to ensure that the discriminative accuracy of sensitivity is above 93.4%. As the model fidelity and number of discriminative indicators increased, the accuracy of the analysis improved significantly, but the analysis accuracy rate of improvement decreased when both increased to a certain level. Therefore, the fidelity of the model used in this study was 5%, and the number of discriminative indicators was three, which met the requirement of the accuracy of the study.

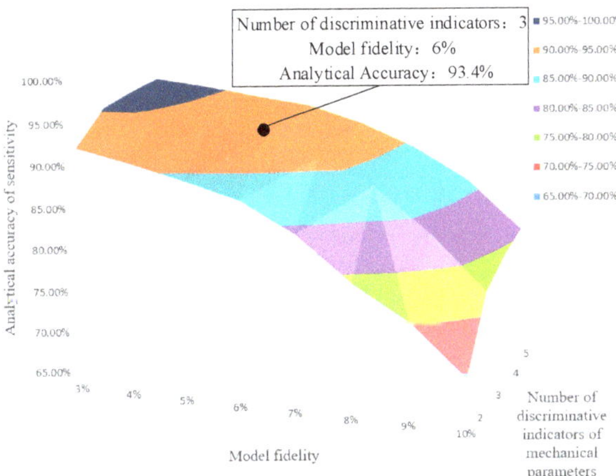

Figure 8. Relationship between the accuracy of sensitive analysis and the fidelity of the model and the number of discriminative indicators.

5. Discussion and Conclusions

DT technology is a central to realizing intelligent analysis of structures. A multidimensional model oriented to the intelligent discrimination of structural sensitivity driven by DTs is established, and an analysis method for the intelligent discrimination of prestressed cable sensitivity is proposed by integrating D–S evidence theory to improve the accuracy of structural safety analysis and provide a reliable basis for the maintenance of the structure. In the present study, the following main findings were obtained.

1. Through the capture of monitored and simulated data, a virtual model of the structure with high fidelity can be established using the average weighting method, which provides a model basis for the intelligent analysis of the structure.
2. On the basis of establishing the twin model, the mechanical parameters of the structure are simulated by setting the working conditions. Further, the mechanical properties of each node can be analyzed with high accuracy by integrating multiclass discriminant indices from D–S evidence theory, which guarantees the safety assessment and accurate maintenance of the structure.
3. Through the integration of DTs and D–S evidence theory, the mapping of virtual space to real tension can be realized, and it also provides a reference for the virtual space to control the real tension.

Driven by the DT-based intelligent discrimination method for the sensitivity of the mechanical parameters of prestressed cables, the safety assessment of structures can be carried out by fusing intelligent algorithms with simulated data from twin models, which effectively improves the intelligence and accuracy of structural analysis and also reduces the cost of real-time monitoring of real structures. This overcomes the misjudgment of structural safety performance owing to errors in the data collected by sensors. The change in structural safety performance is related to influencing factors such as the length error of members, wind load, and temperature. In future research, the influence of the change degree of mechanical parameters on the development trend of structural reliability will be focused on according to the concept of DTs and under the comprehensive action of various factors.

Author Contributions: Conceptualization, Z.L.; methodology, Z.L.; software, Z.L.; validation, Z.L., G.S., A.J., and W.L.; writing—original draft preparation, G.S.; writing, review and editing, Z.L.; project administration, Z.L.; funding acquisition, Z.L. All authors have read and agreed to the published version of the manuscript.

Funding: The research was funded by the National Key R&D Program for the 13th-Five-Year Plan of China (grant number 2018YFF0300300 The research was funded by the Natural Science Foundation of Beijing, Beijing, China (grant number 8202001).

Institutional Review Board Statement: Not applicable.

Informed Consent Statement: Not applicable.

Data Availability Statement: Data sharing not applicable.

Acknowledgments: The authors would like to thank Beijing University of Technology, Beijing, China, for their support throughout the research project.

Conflicts of Interest: The authors declare no conflict of interest.

References

1. Kotsovinos, P.; Judge, R.; Walker, G.; Woodburn, P. Fire Performance of Structural Cables: Current Understanding, Knowledge Gaps, and Proposed Research Agenda. *J. Struct. Eng.* **2020**, *146*, 03120002. [CrossRef]
2. Krishnan, S. Structural design and behavior of prestressed cable domes. *Eng. Struct.* **2020**, *209*, 110294. [CrossRef]
3. Chen, Z.H.; Ma, Q.; Yan, X.Y.; Lou, S.Y.; Chen, R.H.; Si, B. Research on Influence of Construction Error and Controlling Techniques of Compound Cable Dome. *J. Hunan Univ.* **2018**, *45*, 47–56. (In Chinese)
4. Guo, Y.L.; Zhang, X.Q. Experimental study on the influences of cable length errors in Geiger cable dome designed with un-adjustable cable length. *China Civil. Eng. J.* **2018**, *51*, 52–68. (In Chinese)
5. Zhang, A.L.; Sun, C.; Jiang, Z.Q. Calculation Method of Prestress Distribution for Levy Cable Dome with Double Struts Considering Self-weight. *Eng. Mech.* **2017**, *34*, 211–218. (In Chinese)
6. Xue, S.D.; Tian, X.S.; Liu, Y.; Li, X.Y.; Liu, R.J. Mechanical behavior of single-layer saddle-shape crossed cable net without inner-ring. *J. Build. Struct.* **2021**, *42*, 30–38. (In Chinese)
7. Liu, Z.S.; Han, Z.B.; He, J.; Wang, Z.Q. Sensitive Test on Relaxation of Cable and Reliability Assessment of Spoke Cable-truss Structure. *J. Tongji Univ.* **2019**, *47*, 946–956. (In Chinese)
8. Asadolahi, S.M.; Fanaie, N. Performance of self-centering steel moment frame considering stress relaxation in prestressed cables. *Adv. Struct. Eng.* **2020**, *23*, 1813–1822. [CrossRef]
9. Arezki, S.; Kamel, L.; Amar, K. Effects of temperature changes on the behavior of a cable truss system. *J. Constr. Steel Res.* **2017**, *129*, 111–118.
10. Castillo, E.; Ramos, A.; Koller, R.; López-Aenlle, M.; Fernández-Canteli, A. A critical comparison of two models for assessment of atigue data. *Int. J. Fatigue* **2008**, *30*, 45–57. [CrossRef]
11. Shekastehband, B.; Abedi, K.; Dianat, N.; Chenaghlou, M.R. Experimental and numerical studies on the collapse behavior of tensegrity systems considering cable rupture and strut collapse with snap-through. *Int. J. Non Linear Mech.* **2012**, *47*, 751–768. [CrossRef]
12. Liu, Z.; Shi, G.; Zhang, A.; Huang, C. Intelligent Tensioning Method for Prestressed Cables Based on Digital Twins and Artificial Intelligence. *Sensors* **2020**, *20*, 7006. [CrossRef] [PubMed]
13. Greif, T.; Stein, N.; Flath, C.M. Peeking into the void: Digital twins for construction site logistics. *Comput. Ind.* **2020**, *121*, 103264. [CrossRef]
14. Zhang, X.H.; Zhang, C.; Wang, M.Y.; Wang, Y.; Du, Y.Y.; Mao, Q.H.; Lv, X.Y. Digital Twin-driven Virtual Control Technology of Cantilever Roadheader.[J/OL]. *Comput. Integr. Manuf. Syst.* 1–8. Available online: http://kns.cnki.net/kcms/detail/11.5946.TP.20201026.1618.050.html (accessed on 6 February 2021). (In Chinese).
15. Gong, X.Y.; Lei, K.F.; Wu, Q.Y.; Cui, X.Q.; Wu, Y.; Zhu, B.; Yang, F.Q.; Zhang, H.B.; Liu, H. Digital twin driven airflow intelligent control system for the air outlet of the tunneling face. [J/OL]. *China Coal Soc.* **2020**, 1–10. (In Chinese) [CrossRef]
16. Meng, S.H.; Ye, Y.M.; Yang, Q.; Huang, Z.; Xie, W.H. Digital twin and its aerospace applications. *Acta Aeronaut. Astronaut. Sin.* **2020**, *41*, 6–17. (In Chinese)
17. Szpytko, J.; Duarte, Y.S. A digital twins concept model for integrated maintenance: A case study for crane operation. *J. Intell. Manuf.* **2020**, 1–19. [CrossRef]
18. Zhou, Y.; Xing, T.; Song, Y.; Li, Y.; Zhu, X.; Li, G.; Ding, S. Digital-twin-driven geometric optimization of centrifugal impeller with free-form blades for five-axis flank milling. *J. Manuf. Syst.* **2020**. [CrossRef]
19. Yu, J.S.; Song, Y.; Tang, D.Y.; Dai, J. A Digital Twin approach based on nonparametric Bayesian network for complex system health monitoring. *J. Manuf. Syst.* **2020**. [CrossRef]
20. Venkatesan, S.; Manickavasagam, K.; Tengenkai, N.; Vijayalakshmi, N. Health monitoring and prognosis of electric vehicle motor using intelligent-digital twin. *IET Electr. Power Appl.* **2019**, *13*, 1328–1335. [CrossRef]
21. Seon, G.; Nikishkov, Y.; Makeev, A.; Ferguson, L. Towards a Digital Twin for Mitigating Void Formation in Autoclave Composite Parts. *Eng. Fract. Mech.* **2019**, *225*, 106792. [CrossRef]
22. Angjeliu, G.; Coronelli, D.; Cardani, G. Development of the simulation model for Digital Twin applications in historical masonry buildings: The integration between numerical and experimental reality. *Comput. Struct.* **2020**, *238*, 106282. [CrossRef]

23. Hu, Z.Z.; Tian, P.L.; Li, S.W.; Zhang, J.P. BIM-based integrated delivery technologies for intelligent MEP management in the operation and maintenance phase. *Adv. Eng. Softw.* **2018**, *115*, 1–16. [CrossRef]
24. Stepinac, M.; Gašparović, M. A Review of Emerging Technologies for an Assessment of Safety and Seismic Vulnerability and Damage Detection of Existing Masonry Structures. *Appl. Sci.* **2020**, *10*, 5060. [CrossRef]
25. Chen, W.T.; Tsai, I.C.; Merrett, H.C.; Lu, S.T.; Lee, Y.I.; You, J.K.; Mortis, L. Construction Safety Success Factors: A Taiwanese Case Study. *Sustainability* **2020**, *12*, 6326. [CrossRef]
26. Li, G.Q.; Cao, K.; Lu, Y.; Jiang, J. Effective length factor of columns in non-sway modular steel buildings. *Adv. Steel Constr.* **2017**, *13*, 412–426.
27. Lacey, A.W.; Chen, W.; Hao, H.; Bi, K. Effect of inter-module connection stiffness on structural response of a modular steel building subjected to wind and earthquake load. *Eng. Struct.* **2020**, *213*, 110628. [CrossRef]
28. Chen, W.; Ye, J.H.; Jin, L.; Jiang, J.; Liu, K.; Zhang, M.; Chen, W.W.; Zhang, H. High-temperature material degradation of Q345 cold-formed steel during full-range compartment fires. *J. Constr. Steel Res.* **2020**, *175*, 106366. [CrossRef]
29. Okazaki, S.; Okuma, C.; Kurumatani, M.; Yoshida, H.; Matsushima, M. Predicting the Width of Corrosion-Induced Cracks in Reinforced Concrete Using a Damage Model Based on Fracture Mechanics. *Appl. Sci.* **2020**, *10*, 5272. [CrossRef]
30. Zeng, B.; Zhou, Z.; Zhang, Q.F.; Xu, Q.; Meng, S.P. Analytical and experimental research on damage identification of cable-stayed arch-truss based on data fusion. *China Civil. Eng. J.* **2020**, *53*, 28–37+86. (In Chinese)
31. Bappy, M.M.; Ali, S.M.; Kabir, G.; Paul, S.K. Supply chain sustainability assessment with Dempster-Shafer evidence theory: Implications in cleaner production. *J. Clean. Prod.* **2019**, *237*, 117771. [CrossRef]
32. Denoeux, T.; Shenoy, P. An Interval-Valued Utility Theory for Decision Making with Dempster-Shafer Belief Functions. *Int. J. Approx. Reason.* **2020**, *124*, 194–216. [CrossRef]

Article

Application of Analytical Hierarchy Process for Structural Health Monitoring and Prioritizing Concrete Bridges in Iran

Saeid Darban [1], Hosein Ghasemzadeh Tehrani [2], Nader Karballaeezadeh [2] and Amir Mosavi [3,*]

[1] Department of Engineering, Azad University of Shahrood, Shahrood P.O. Box 3619943189, Iran; Saeid.d1366@yahoo.com
[2] Faculty of Civil Engineering, Shahrood University of Technology, Shahrood P.O. Box 3619995161, Iran; H_ghasemzadeh@shahroodut.ac.ir (H.G.T.); N.karballaeezadeh@shahroodut.ac.ir (N.K.)
[3] John von Neumann Faculty of Informatics, Obuda University, 1034 Budapest, Hungary
* Correspondence: amir.mosavi@mailbox.tu-dresden.de

Citation: Darban, S.; Ghasemzadeh Tehrani, H.; Karballaeezadeh, N.; Mosavi, A. Application of Analytical Hierarchy Process for Structural Health Monitoring and Prioritizing Concrete Bridges in Iran. *Appl. Sci.* **2021**, *11*, 8060. https://doi.org/10.3390/app11178060

Academic Editors: Hugo Filipe Pinheiro Rodrigues and Ivan Duvnjak

Received: 29 March 2021
Accepted: 22 May 2021
Published: 31 August 2021

Publisher's Note: MDPI stays neutral with regard to jurisdictional claims in published maps and institutional affiliations.

Copyright: © 2021 by the authors. Licensee MDPI, Basel, Switzerland. This article is an open access article distributed under the terms and conditions of the Creative Commons Attribution (CC BY) license (https://creativecommons.org/licenses/by/4.0/).

Abstract: This paper proposes a method for monitoring the structural health of concrete bridges in Iran. In this method, the bridge condition index (BCI) of bridges is determined by the analytical hierarchy process (AHP). BCI constitutes eight indices that are scored based on the experts' views, including structural, hydrology and climate, safety, load impact, geotechnical and seismicity, strategic importance, facilities, and traffic and pavement. Experts' views were analyzed by Expert Choice software, and the relative importance (weight) of all eight indices were determined using AHP. Moreover, the scores of indices for various conditions were extracted from experts' standpoints. BCI defines as the sum of weighted scores of indices. Bridge inspectors can examine the bridge, determine the scores of indices, and compute BCI. Higher values of BCI indicate better conditions. Therefore, bridges with lower BCI take priority in maintenance activities. As the case studies, the authors selected five bridges in Iran. Successful implementation of the proposed method for these case studies verified that this method can be applied as an easy-to-use optimization tool in health monitoring and prioritizing programs.

Keywords: transportation infrastructure; concrete bridges; structural health monitoring; bridge condition index; analytical hierarchy process; mobility; multiple-criteria decision analysis; decision making; civil engineering; infrastructure

1. Introduction

The quality of transportation systems directly affects the lives of urban residents. A large portion of the national resources of each country is invested in this area. As one of the most important parts of transportation systems, bridges have a critical role in urban development [1–5]. The bridge conditions in the transportation networks are so important that the costs incurred by out-of-service bridges are exorbitant. Therefore, bridge condition evaluation has crucial importance for the proper maintenance and management of transportation infrastructures. Another important factor that affects the maintenance process of infrastructures is budget constraints. Consequently, further attention should be paid to the development of a bridge management system (BMS) [6,7]. The first step in the BMS is to prepare a technical profile for all bridges in the network. This profile contains technical information such as the name of a bridge, its location, construction method, etc. It is, in fact, the starting point of BMS. The next step in BMS is assessment, including structural and seismic assessment, hydrological assessment, facility evaluation, safety assessment, and pavement and traffic evaluation [8]. Bridge inspection methods are divided into four general categories [9,10]: 1. Visual assessment; 2. Evaluation by non-destructive tests; 3. Sampling and destructive tests; and 4. Health assessment. Another major step of BMS is bridge maintenance. The maintenance involves a variety of operations that continuously ensure the safety and serviceability of bridges over their lifetime. The prioritization of bridge maintenance, including repairs or reinforcement, is the cornerstone of the BMS [8,11]. Traditionally, for small-sized bridges, the prioritization of bridge maintenance projects was carried out based on engineer's assessments. In large and old bridge networks, it was conducted in accordance with concepts and principles of optimization in project budget allocation. Today, the bridge condition index (BCI) is used for this purpose. BCI is a good benchmark for prioritizing BMS [12].

The service life of a bridge is divided into four different phases [13]:

- Design and construction;
- Start of damages (early damage stages);
- The spread of damages;
- The expansion of damages.

Under the famous Law of Five, each dollar spent on the first phase will equal $5 in the second phase, $25 in the third phase, and $125 in the fourth phase [13]. According to this law, any miscalculated decisions about maintenance, repair, and rehabilitation (MR&R) in bridges would incur surplus costs. With this in mind, there is a need for a decision support system (DSS). DDS aims at improving the bridge network condition and allocating the budget appropriately [14]. Most of BMSs are founded upon processes that optimize the cost of a lifecycle. They tend to overlook factors such as environmental impacts and social impacts. This gives rise to a number of problems, especially when the existing financial resources are higher or lower than the cost of the computational life cycle [15].

In this paper, the main goal is to present an applicable method for determining the condition index of the concrete bridges in Iran. For this purpose, firstly, eight critical indices were selected. These indices include structure, hydrology and climate, safety, load impact, geotechnical and seismicity, strategic importance, facilities, and traffic and pavement. Each index comprises a number of sub-indices. Next, the authors developed a questionnaire for surveying the views of experts. The questionnaire was about the relative importance of the indices and sub-indices. Moreover, examining various conditions of sub-indexes is another aim of this questionnaire. After that, this questionnaire was distributed among bridge experts. Then, experts' feedbacks about the relative importance of indices and some sub-indexes were analyzed by Expert Choice software. This software resulted in the relative weight of these indexes and sub-indexes using the analytical hierarchy process (AHP). For all sub-indexes, experts' views about conditions scores were gathered as well. The score of each index is the sum of condition scores assigned to its sub-indices. Finally, BCI is calculated as the sum of weighted scores assigned to indices. BCI is a value between 0 and 100, with higher values indicating a better bridge condition. Therefore, bridges with lower BCI take priority in terms of repair and maintenance. To test the proposed method in practice, five bridges in Semnan province, in Iran, were inspected and their BCI was determined to prioritize bridges in terms of maintenance requirements. This study is innovative because no comprehensive method has been proposed to evaluate and prioritize in-service bridges in Iran. Therefore, the proposed method helps Iranian engineers evaluate bridges and prioritize bridge maintenance operations more effectively.

This paper is organized as follows: Next section presents a literature review of BCI. Section "Method" introduces the study methodology, which is further divided into three general sub-sections entitled BCI, AHP, and sub-indices of BCI. The results are presented and discussed in section "Results and Discussion." The final section offers a summary of results and conclusions.

2. Background

The proper maintenance and management of bridges need the evaluation of safety and lifetime conditions. In recent years, there is an increasing number of studies on the BMS and BCI. This section presents a range of the most important methods used for determining BCI around the world.

2.1. India

In India, Sanjay and Kumar developed a bridge health index (BHI) using AHP. They divided elements of the bridge into seven categories, including approaches, substructure, waterway/channel, foundations, superstructure, appurtenances/auxiliary works, and bearings. Then, they drafted a questionnaire and distributed it among engineers and experts. The results of the questionnaire were incorporated in determining the relative importance and weight of diverse elements. They also considered a numerical value for each type of damage. The condition of various elements of a bridge is assessed by visual inspection. Finally, BHI was developed by summing the score of all bridge elements [16]. In Figure 1, the decision tree of this research is presented.

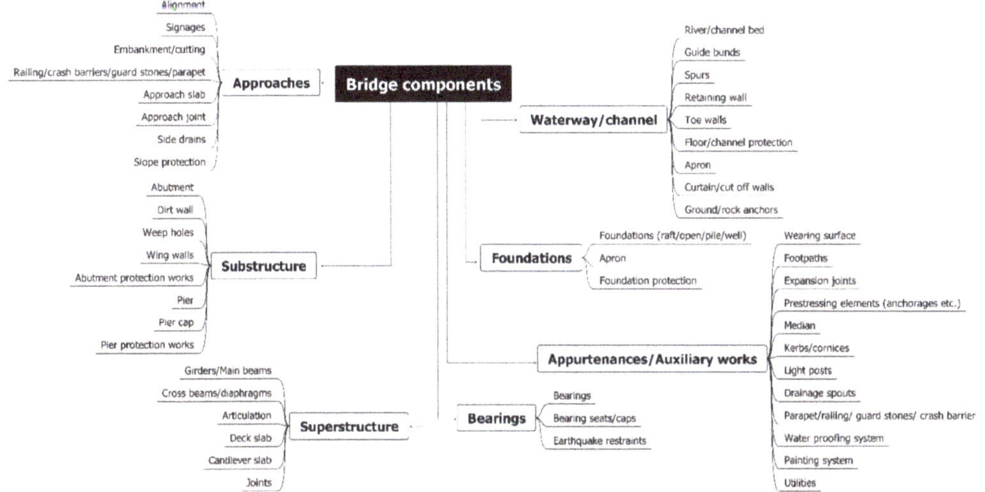

Figure 1. Decision tree in Sanjay and Kumar research.

2.2. China

In China, there are two main indexes for assessing bridge conditions. The Ministry of Transport of the People's Republic of China uses the Dr index to assess the conditions of a bridge [17]:

$$D_r = BDCI \times W_D + SPCI \times W_{SP} + SBCI \times W_{SB}, \tag{1}$$

where D_r is bridge condition rating, BDCI is bridge deck condition index, SPCI is bridge superstructure condition index, SBCI is bridge substructure condition index, W_D, W_{SP}, and W_{SB} are the weight of BDCI, SPCI, and SBCI indicators, respectively. On the other hand, the Ministry of Housing and Urban-Rural Development of China has also provided a definition identical to Equation (1) for assessing bridge conditions [18]:

$$BCI = BCI_d \times \omega_d + BCI_{sp} \times \omega_{sp} + BCI_{sb} \times \omega_{sb}, \tag{2}$$

where BCI is bridge conditions index, BCI_d is bridge deck condition index, BCI_{sp} is bridge superstructure condition index, BCI_{sb} is bridge substructure condition index, and ω is the weight of a bridge element. Table 1 presents the assessment approach based on D_r and BCI.

Table 1. Assessment of bridge condition in China.

D_r		Condition
$95 \leq D_r < 100$	$90 \leq BCI$	Perfect
$80 \leq D_r < 95$	$80 \leq BCI < 90$	Good (minor damage)
$60 \leq D_r < 80$	$66 \leq BCI < 80$	Pass (mediate damage)
$40 \leq D_r < 60$	$50 \leq BCI < 66$	Unqualified (great damage)
$0 < D_r < 40$	$BCI < 50$	Dangerous

2.3. Japan

In Japan, no specific formula or equation is used to evaluate the condition of a bridge. For each bridge, first, one of the statuses shown in Table 2 is assigned to each element based on the assessment of the bridge inspector, and then the bridge general conditions are described in accordance with Table 3 [19].

Table 2. Assessment of maintenance urgency for bridge element in Japan.

Rate	Condition
A	No repair needed
B	No immediate repairs needed
C1	Immediate repairs needed from standpoint of preventative maintenance
C2	Immediate repairs needed from standpoint of structural safety
E1	Immediate actions needed from standpoint of structural safety
E2	Immediate actions needed in tandem with other factors
M	Repairs needed during regular maintenance work
S1	In-depth investigations needed
S2	Follow-up investigations needed

Table 3. Bridge soundness in Japan.

State	Condition	Description
1	Good	No problems in bridge's functions
2	Preventative maintenance	No problems in bridge's functions but maintenance required from standpoint of preventive maintenance
3	Early action	Possibility of problems in bridge's functions, need for early action
4	Emergency action	Possibility of problems or existing problems in bridge's functions, need for emergency actions

2.4. Korea

In Korea, a damage index (DI) is used to assess bridge conditions. It is the normalized index obtained from the evaluation of all bridge elements. The DI index is shown in Equation (3) [20]:

$$DI = \sum (CR_i \times WF_i)/100, \tag{3}$$

$$\sum (WF_i) = 100,$$

where DI is damage index, CR_i is condition evaluation of ith element, and WF_i is the weight factor of ith element. Based on the DI index, a bridge condition could be described with grades A to E (Table 4).

Table 4. Assessment of bridge condition in Korea.

Rate	DI	Description
A	$0 \leq DI < 0.13$	Perfect
B	$0.13 \leq DI < 0.26$	Minor problem in secondary elements
C	$0.26 \leq DI < 0.49$	Minor problem in primary elements
D	$0.49 \leq DI < 0.79$	Problem in primary elements
E	$0.79 \leq DI$	Serious problem in primary elements

2.5. United States

In the United States, there are various approaches to assess the condition of the bridge. For example, the California Department of Transportation defines BHI based on Equation (4). This index varies from 0 for the worst bridge condition to 100 for the healthiest bridge condition [21].

$$BHI = [\textstyle\sum(CEV)/\sum(TEV)] \times 100, \tag{4}$$

where BHI is bridge health index, CEV is current element value, and TEV is total element value.

CEV and TEV can be calculated according to the following equations [21]:

$$CEV = \textstyle\sum(QCS_i \times WF_i) \times FC, \tag{5}$$

$$TEV = TEQ \times FC, \tag{6}$$

where TEQ is total element quantity, FC is failure costs of element, QCS_i is quantity in condition state i, and WF is weight factor.

In the United States, transportation departments report a set of data called national bridge inspection (NBI). Based on the physical condition of the bridge, the bridge is assigned a score in the range of 0 to 9 [22]. The assessment procedure is presented in Table 5.

Table 5. Assessment of bridge condition based on NBI.

Rate	State	Description
9	Excellent	A new bridge
8	Very good	No problems noted
7	Good	Some minor problems
6	Satisfactory	Structural elements show some minor deterioration
5	Fair	All primary structural elements are sound but may have minor section loss, deterioration, spalling or scour
4	Poor	Advanced section loss, deterioration, spalling, scour
3	Serious	Loss of section, etc., has affected primary structural components; Local failures are possible. Fatigue cracks in steel or shear cracks in concrete may be present
2	Critical	Advanced deterioration of primary structural elements. Fatigue cracks in steel or shear cracks in concrete may be present or scour may have removed structural support. Unless closely monitored it may be necessary to close the bridge until corrective action is taken
1	Imminent failure	Major deterioration or loss of section in critical structural components or obvious vertical or horizontal movement affecting structural stability. Bridge is closed to traffic but corrective action may allow it to be returned to light service
0	Failed	Out of service. Beyond corrective action

The US departments of transportation often use a computer program to assess bridge conditions. This program is based on Equation (7) [22].

$$SR = S_1 + S_2 + S_3 - S_4, \tag{7}$$

where SR is Sufficiency rating, S_1 is the parameter related to structural safety, S_2 is the parameter related to bridge serviceability and functionality, S_3 is the parameter related to user requirements, and S_4 is the parameter related to reductive coefficients based on structure type and traffic safety.

SR indicates the bridge sufficiency to remain in service, where SR has a maximum rating of 100%, indicating complete bridge sufficiency, and a minimum rating of 0%, indicating complete bridge deficiency. The parameters S_1, S_2, S_3, and S_4 have weight importance of 55%, 30%, 15%, and 13%, respectively. FHWA uses SR to allocate rebuilding funds so that [22]:

- If SR < 50, the bridge is eligible for replacement;
- If 50 < SR < 80, the bridge is eligible for rehabilitation.

2.6. Australia

In Australia, Rashidi et al. presented an overall working framework for bridge infrastructure management. This framework had two phases, including project ranking and remediation planning. The engaged factors of this framework were weighting by expert judgments and employing AHP [23]. In phase project ranking, they presented a model for prioritizing based on priority index (PI). A bridge with higher PI takes priority for maintenance [24]:

$$PI = 0.6(SE) + 0.2(FE) + 0.2(CIF), \tag{8}$$

where PI is priority index, SE is structural efficiency index, FE is functional efficiency, and CIF is the client impact factor.

In phase remediation planning, the problem was modeled in a hierarchical order by a simplified hierarchical analysis process (S-AHP). This hierarchy consists of at least three main levels: goal, criteria, and alternatives. The goal is the remediation strategy. The criteria require to be broken down into more specific sub-criteria introduced as attributes in an extra level of the hierarchy. Each criterion has a weight indicating its importance. These weights are defined by the decision makers. The final level is added for the remediation treatment alternatives. This procedure is flexible and can vary for different projects. Therefore, criteria, rehabilitation strategies, and even the number of levels can be different in various cases [25,26].

2.7. Turkey

In Turkey, the following technique is used for the assessment of bridge condition and its elements [27]:

$$CR\ (e)_{element,W} = \sum j \left(\left(\sum (WP_{dt,i,j} \times r_j) \right) / \left(\sum\sum (WP_{dt,i,n} \times r_n) \right) \right), j = 1, \dots, S; i = 1, \dots, d; \\ n = 1, \dots, S \tag{9}$$

$$CR\ (b)_{bridge,W} = \sum (W_e \times CR\ (e)_{element,W}) / 100, e = 1, \dots, n_e \tag{10}$$

where $CR(e)_{element,W}$ is weighted assessment of conditions for element e, n_e is total number of bridge elements, $WP_{(dt,i,j)}$ is weighted percentages for damage type i under condition j, s is total conditions, d is total number of damages, r_j is damage impact distribution coefficient, $CR(b)_{bridge,W}$ is weighted assessment of bridge b, and W_e is weight importance of r_j element. In Figure 2, the weight importance of elements for a conventional concrete bridge is shown.

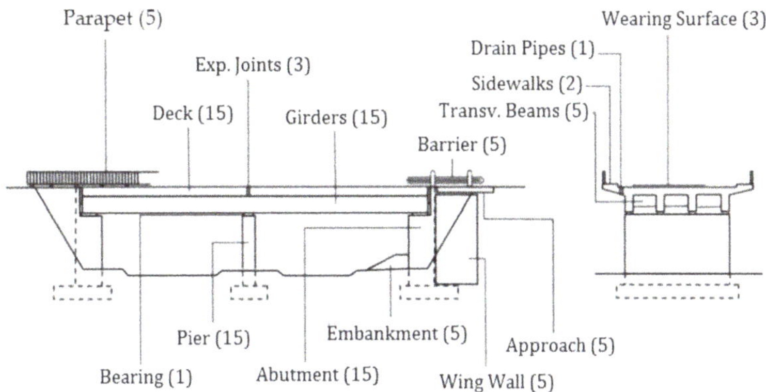

Figure 2. Weight importance of elements for a conventional concrete bridge.

2.8. Concluding Remarks

In previous parts of this section, an overview of the most important studies is relating to methods of assessing the condition of bridges in different countries is presented. After reviewing these works, the authors found that the main gap in these methods is: almost all methods/models only focus on the structural condition of bridges. Of course, considering the structural condition is logical and essential because the structural elements play the most key role in bridge serviceability. However, disregarding other important factors, including hydrology, climate, safety, load impact, etc., can decrease the quality of monitoring, assessing, and prioritizing bridges. Another key point that can be found from the background is the attempt of some researchers and organizations to resolve the mentioned gap. By applying some new factors such as safety, seismic evaluation, and hydrology, they tried to remove/moderate the impact of the gap. This is a valuable attempt and a helpful step in eliminating the mentioned gap. But, the authors think that these attempts can be more extended and with more details. As a result, the authors aim to present a new method that includes a wide range of the affecting factors in evaluating bridges condition. Indeed, this study presents a method for monitoring, assessing, and prioritizing bridges that is not confined to structural condition and considers other affecting factors. Moreover, all these factors are examined in detail as much as possible. The method includes hydrology and climate index, safety index, load impact index, geotechnical and seismic index, mechanical/electrical facilities quality index, strategic importance index, and traffic and pavement index. Another gap that has been followed in this study is the lack of a comprehensive, practical, scientific method for bridge networks in Iran. Although there are attempts in the field of developing health monitoring methods for various infrastructures in Iran, such as [28–37]; however, the absence of a full detailed, efficient, specialized method for bridges can be felt. Consequently, the authors also tried to fill the later gap, and therefore, the proposed method is based on the condition of bridges in Iran.

3. Method

3.1. Bridge Condition Index (BCI)

One of the major concerns of organizations in charge of bridges is that repairs and maintenance of bridges should be implemented with respect to financial constraints. Bridge maintenance is a costly and long-term project, which has led to the development of various scientific tools and methods for optimal budget allocation [38]. Before allocating any budget, it is necessary to determine the current condition of the bridge and its present and possible future needs. The key to the successful assessment of a bridge condition is to recognize various damages. Bridge damage is a slow, progressive, and continuous process that is influenced by the imposing load, conditions of various bridge components, environmental factors, and the properties of materials [16].

Many researchers have argued that the damage process is a blend of several mechanisms, such as corrosion, creep, shrinkage, cracking, fatigue, etc. [39]. The bridge damage is induced by a host of factors such as traffic, rainfall, freezing and melting cycles, climate change, and pollution, which can eventually lead to bridge failure [16]. The bridge failure can be either structural or functional. The methods for assessing various components of a bridge and their relative significance are key concepts in BMS [40].

Different countries employ diverse methods to evaluate bridges so that they can develop a priority plan for bridge repair and maintenance with respect to budget constraints. Using a series of indicators is one of the most commonly used decision-making methods to prioritize maintenance. One perquisite of such indicators is determining the relative importance of different bridge components. Indicators can be categorized into two broad categories [16]:

1. Bridge Health Index (BHI);
2. Maintenance Priority Index (MPI).

The BHI is generally calculated as follows [16]:

$$\text{BHI} = \sum W_i \times C_i, i = 1, \dots, n \tag{11}$$

where BHI is bridge health index, W_i is the weight of ith element, C_i is the condition of ith element, and n is the number of bridge elements.

Moreover, MPI is usually calculated using the following equation [41]:

$$\text{MPI} = \sum K_i \times F_i(a, b, c, \dots), \tag{12}$$

where MPI is maintenance priority index, K_i is the weight of ith damage, F_i is ith damage and a, b, c, etc., are damage characteristics.

The above indicators, BHI and MPI, aim to determine the condition of the in-service bridges. In fact, both these indicators are somehow the same as the BCI. In this study, a method of BCI determination is proposed that is structurally similar to BHI.

3.2. Analytical Hierarchy Process (AHP)

Introduced by Thomas Satty in 1980, the analytic hierarchy process (AHP) provides a mechanism for switching the criteria rating into weights [4,42]. AHP is an effective and powerful tool for multi-criteria decision-making approaches. In fact, it is a powerful technique for solving complicated problems that may have correlations and interactions among different goals. When using in multi-criteria problems, AHP breaks down these problems into multiple levels of hierarchy. The goal or objective seats in the top level. Intermediate levels include the criteria and sub-criteria, and the lowest level can provide alternatives. AHP then develops priorities among all the criteria and sub-criteria within each level of the hierarchy [43,44]. The basis of AHP can be both experts' judgments and predetermined measurements. Experts are interviewed, and pair-wise comparison judgments are applied to pairs of criteria. Eventually, priorities will be determined. AHP is easy to apply and helps engineers obtain the final ranking from the individual evaluations, and finally, select an optimal alternative [45]. Because of these features, AHP has been used in bridge engineering in the past decades. Generally, the AHP can be included in the following steps [46]:

1. Constructing the pair-wise comparison judgment matrix.
2. Determining the weight of decision elements.
3. Controlling the compatibility index.

The purpose of this study is to present a simple, applicable methodology for the health monitoring and prioritizing of bridges. This methodology works based on calculating the bridge condition index (BCI). For the determination of BCI, the methodology uses AHP. In cases where AHP is used for decision making, a proper hierarchy tree should be designed. This hierarchy tree has different levels depending on the type of problem under investigation. As mentioned earlier, the first level of this tree is goal or objective. In this study, the goal is to determine the BCI of the bridge. In intermediate levels, first, criteria are located. These criteria are eight indices affecting the quality of a bridge, including structural index, hydrology and climate index, safety index, bridge performance index (load impact), geotechnical and seismic index, strategic importance index, facilities index, and traffic and pavement index. Sub-criteria are set in the next level. The sub-criteria, in fact, are sub-indexes of eight indexes of the previous level. Figure 3 shows the hierarchy tree of this research.

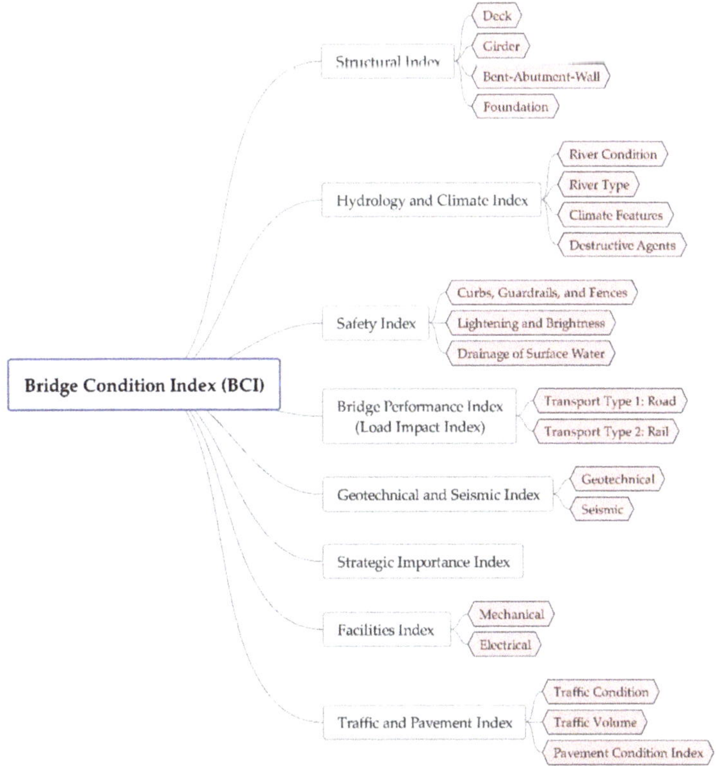

Figure 3. Hierarchy tree in this study.

In this work, the problem is to calculate BCI. For solving this problem, the authors must provide the relative importance of all indexes and sub-indexes. The relative importance values depend on weights which are measured using the AHP method. AHP is an analytical method that allows making appropriate decisions by considering qualitative, quantitative, and mixed criteria. This process is based on a dual comparison system [47]. For this reason, a questionnaire (Appendix A) is designed and distributed among experts. The experts' group comprises several specialists and advisors involved in the maintenance and reinforcement of bridges, also, university professors with relevant expertise. In this questionnaire, firstly, experts are asked to determine the relative importance of eight indices of Figure 3. Table A1 was designed for this purpose. In each cell of Table A1, experts assigned a value between 0 and 10 based on their technical experience and expertise. In fact, this table is a dual comparison between all indices. In this table, if a row index outweighs a column index, experts assigned a value between 1 and 10. If the column index outweighs the row index, a value between 0 and 1 was assigned by experts. Of course, the value of 1 was used at the diameter of the table. Table 6 shows the final results. In fact, the numbers in Table 6 are the average of experts' viewpoints. After this step, the values of Table 6 were entered in Expert Choice software, and the relative weights of all eight indexes were calculated (Table 7).

Table 6. Relative importance of indices based on a survey of experts.

Index	Structural	Hydrology and Climate	Safety	Bridge Performance (Load Impact)	Geotechnical and Seismic	Strategic Importance	Facilities	Traffic and Pavement
Structural	1	5.271	3.152	4.581	1.877	3.13	6.075	3.578
Hydrology and Climate		1	1.037	1.382	0.788	1.377	1.871	0.941
Safety			1	2.613	1.633	1.489	3.318	2.074
Bridge Performance (load impact)				1	0.761	1.164	2	1.154
Geotechnical and Seismic					1	2.859	3.133	2.216
Strategic Importance						1	2.766	1.75
Facilities							1	0.975
Traffic and Pavement								1

Table 7. Relative weights of indices and compatibility rating.

Index	Structural	Hydrology and Climate	Safety	Bridge Performance (Load Impact)	Geotechnical and Seismic	Strategic Importance	Facilities	Traffic and Pavement
Relative weight	0.331	0.097	0.146	0.080	0.143	0.088	0.046	0.068
compatibility rating				0.03				

As indicated in Table 7, the relative weight of indices was calculated. This table contains additional information called compatibility rating. It is the mechanism that determines the adaptability of comparisons, indicating the extent to which the priorities selected by the group or the priorities of the mixed table are reliable. According to the experience, if the compatibility rate is less than 0.1, the adaptability of comparisons is acceptable; otherwise, the comparisons need be repeated [46].

The next step is to examine all indices in detail. In Figure 3, each index is divided into some sub-indices. In this step, the relative importance/weight of all sub-indices must be determined. Figure 3 shows that the structural index consists of four sub-indices. These four sub-indices have various relative importance. Therefore, Table A2 in the questionnaire was assigned to the relative importance of these four sub-indices. In other indices of Figure 3, all sub-indices have the same importance/weight. By determining the relative importance/weight of all indices and sub-indices, it is time to determine the score of each sub-index in various conditions. Tables A3–A10 were designed for this purpose. In these tables, experts rated sub-indices in different states. Considering the given explanations, they filled the blank cells of the tables with a number in the range of 0 to 100. Details of this step are presented in the next sections of the paper. Figure 4 depicts the flow chart of AHP in this study.

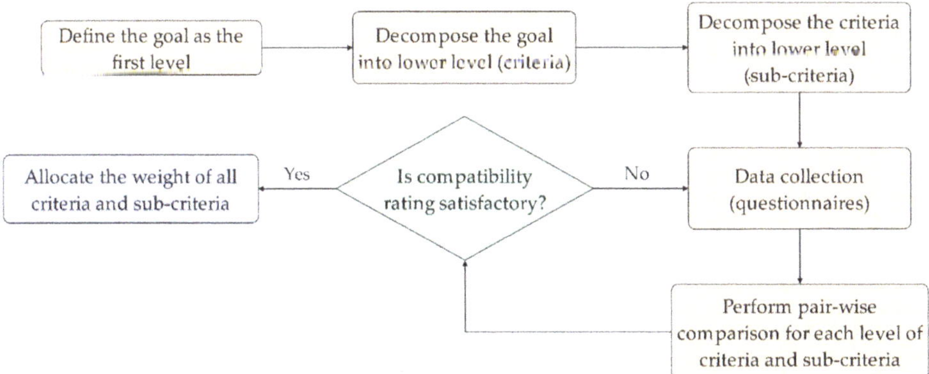

Figure 4. The flow chart of AHP used in this study.

Now, the BCI can be calculated. For each bridge, first, the score of each sub-index is determined based on the bridge inspection. After that, the score of each index is computed as the sum of weighted scores of its sub-indices. Finally, BCI can be determined based on Equation (13):

$$BCI = \sum (X_i \times W_i), i = 1, \dots, 8 \tag{13}$$

where BCI is the bridge condition index, X_i is the score of ith index, and W_i is the weight of ith index.

3.3. Sub-Indices of BCI

3.3.1. Structural Index

The structural index describes the bridge condition in terms of the damages in the structural elements. In other words, it evaluates the structural condition of the bridge. Figure 4 shows that the structural index consists of four sub-indices, including deck, girder, bent-abutment-wall, and foundation. It should be noted that the joints are classified under the deck sub-index and bearing and support in the sub-index of bent-abutment-wall. These four sub-indices have various relative importance. Table A2 in the questionnaire was designed for determining the relative importance of these four sub-indices. Similar to Table A1, the experts were asked to conduct a dual comparison between the four sub-indices and assign a value between 0 and 10 in cells of Table A2. Table 8 presents the averaged viewpoints of experts. By inserting these values into Expert Choice software, the authors calculated the relative weights of the sub-indices. Table 9 indicates these relative weights. Next, these four sub-indices must be assessed for different states of damages. Table A3 had been designed for this purpose. Experts were asked to enter the appropriate scores in the blank cells based on their views about the various damage intensity in each sub-index. Their scores were a number in the range of 0 to 100. A score of 100 is related to the best condition, and a score of 0 is related to the worst condition. Table 10 shows the assigned score of each sub-index for different conditions. These scores are the average of experts' viewpoints.

Table 8. Relative importance of sub-indices in structural index.

Sub-Index	Deck	Girder	Bent-Abutment-Wall	Foundation
Deck	1	1	0.84	1.476
Girder		1	1.644	1.94
Bent-abutment-wall			1	3.204
Foundation				1

Table 9. Relative weight of sub-indices in structural index (Results of Expert Choice software).

Sub-Index	Deck	Girder	Bent-Abutment-Wall	Foundation	Compatibility Rating
Relative weight	0.247	0.32	0.297	0.136	0.04

Table 10. Scores of sub-indexes in structural index.

Damage Intensity	Sub-Indexes Scores			
	Deck	Girder	Bent-Abutment-Wall	Foundation
Low	95	95	90	95
Mediate	70	65	60	75
High	30	30	25	35

The structural index score is the sum of weighted scores of its four sub-indices. In the bridge maintenance program, usually, more attention was allocated to structural sub-indices, and they were included with more details in the assessment. In other indices, however, evaluations were more general.

3.3.2. Hydrology and Climate Index

One of the affecting factors in the condition of a bridge is hydrology and environmental factors. The hydrology and climate index focuses on four factors, including river conditions, river type, climatic features, and the concentration of destructive matters (such as sulfates) in the water, soil, and air. Based on the explanation of Table A4, experts provided the appropriate scores in the range of 0 to 100 for various conditions of this table. A score of 100 is related to the best conditions, and a score of 0 is related to the worst conditions. These scores are shown in Table 11. The score of hydrology and climate index is the average value of these scores, with identical weight. It should be noted that if there is no river in the path under the bridge, the sub-indices of river conditions and destructive matters would be removed.

3.3.3. Safety Index

This index includes parameters that affect the safety of the bridge. These parameters include the beauty and proper serviceability of curbs, absence of crack/fracture/delaminated curbs, proper functioning of the guardrails/fences, lighting and brightness, and the efficiency of the drainage system to provide sufficient friction coefficient. Table A5 in the questionnaire was designated for the safety index. Based on the explanation of Table A5, experts were asked to assign the relevant score for various conditions of safety equipment. These scores are in the range of 0 to 100. Again, the scores of 100 and 0 are related to the best and worst conditions, respectively. Table 12 shows the average scores based on the experts' scores. The safety index score is the average value of scores of these three sub-indexes, with identical weight.

Table 11. Scores of sub-indexes in hydrology and climate index.

Sub-Indexes Scores			
River Conditions		**River Type**	
Description	**Score**	**Type**	**Score**
There is no erosion in the riverbed or the erosion is trivial. The amount of sedimentation and debris is negligible	98	Area under the bridge is not a river path	98
The riverbed has eroded slightly. There are signs of depositions in the upstream and downstream. Further analysis is required to detect failures	58	There is seasonal river flowing under the bridge.	59
The erosion of the riverbed is critical and concerning. There are enormous amounts of sedimentations around the bridge. Serious measures have to be taken.	14	There is permanent river flowing under the bridge.	8
Climatic Features		**Destructive Agents**	
Description	**Score**	**Quality of Protection against Destructive Matters**	**Score**
Mild (there are no invasive agents such as moisture, transpiration, freezing and melting cycle, corrosive substances, etc.)	93	Very good	95
Medium (conditions that are occasionally exposed to moisture and transpiration, and elements that are permanently exposed to non-invasive soils and water, or underwater with a pH > 5)	80	Good	76
Severe (extreme humidity or transpiration, or freezing and thawing cycle, elements immersed in water, such that one surface is exposed to air, elements in chlorine ion air, elements exposed to corrosion caused by the use of anti-freezing agents)	54	Medium	49
Extremely severe (conditions that are exposed to gases, water and static sewage with a pH of up to 5, corrosive matters, moisture with extreme icing and melting)	35	Bad	14
Exceptionally severe (conditions subject to extreme erosion, flowing water and sewage with a maximum pH of 5)	20	-	-

Table 12. Score of sub-indexes in safety index.

Sub-Indexes Scores					
Curbs, Guardrails and Fences		**Lighting and Brightness**		**Drainage of Surface Water**	
Description of Defects	**Score**	**Conditions**	**Score**	**Drainage Condition**	**Score**
No repair is needed	98	Trivial dazzling, excellent color rendering, broad sight	94	Perfect drainage, adequate friction coefficient	96
Partial repair is needed	67	Slight dazzling, color rendering and sight are relatively desirable	66	Drainage for securing desirable friction	68
Major repair is required	14	Extreme dazzling, low color rendering and limited sight	23	Improper drainage, undesirable friction coefficient	27

3.3.4. Load Impact Index

In bridges that are under heavy loads or dynamic loads, damages are more common. The dynamic load imposed on a railway bridge is higher than that of a road bridge. Moreover, bridges for which use crossing heavy vehicles, such as a trailer or trucks, are more likely to be damaged than bridges used for light traffic. These points are considered in the impact load index. Table A6 in the questionnaire was designated for this index. In this table, experts assigned their scores for various classes and types of the transport system that the under-investigation bridge belongs to it. These scores are in the range of 0 to 100. Table 13 shows the final results of Table A6 that is the average value of experts' scores.

Table 13. Load impact score.

Class	Transport Type	
	Car	Train
Freeway	40	30
Highway and major road	45	40
Minor road	70	60
Rural road	85	-
Metro and monorail	-	70

3.3.5. Geotechnical and Seismic Index

The quality of soil under the bridge foundation, seismicity of the region, and its geological structure affect the behavior of the bridge during an earthquake and its settlement, which consequently affect the bridge condition. For considering these points, Table A7 was designed to determine the geotechnical and seismic index. According to the earth and the seismic area type, experts filled the blank cells with numbers from the range of 0 to 100. The better conditions take higher scores and vice versa. The average values of Table A7 are presented in Table 14. Based on Table 14, each bridge takes two scores: Geotechnical score and seismic score. The score of the geotechnical and seismic index is obtained by averaging these two scores.

Table 14. Scores of sub-indexes in geotechnical and seismic index.

Sub-Indexes Scores			
Geotechnical		Seismic	
Earth Type	Score	Seismic Area Type	Score
I	92	Low relative risk	80
II	71	Medium relative risk	63
III	47	High relative risk	40
IV	26	Very high relative risk	23

3.3.6. Strategic Importance Index

This index indicates the importance of the bridge location in terms of regional, strategic, and political considerations. Strategic areas include hospitals (with more than 500 beds), military centers, crisis management centers, and fire stations. The experts were asked to write their scores in Table A8. They used the numbers from the range of 0 to 100. Table 15 indicates the final scores that are the average values of experts' scores.

Table 15. Scores of the strategic importance index.

The Strategic Importance of Bridge	Score
High importance (links two strategic areas)	89
Medium importance (links streets and non-strategic arterial)	55
Low importance (other bridges)	29

3.3.7. Facilities Index

This index is composed of two parts, including mechanical facilities and electrical facilities. The facilities index demonstrates the need for repairing the electrical or mechanical facilities of the bridge. Table A9 in the questionnaire was designed for this index. Experts scored various conditions of these facilities from the range of 0 to 100. Scores of 100 and 0 are related to the best and the worst conditions, respectively. The average values of these scores are presented in Table 16. It is important to point out that both sub-indexes of Table 16 have the same importance. Therefore, the overall score of the facilities index concludes from averaging of sub-indexes scores with equal weights.

Table 16. Scores of sub-indexes in facilities index.

Sub-Indexes Scores			
Mechanical Facilities		**Electrical Facilities**	
Drainage System	**Score**	**Lighting Condition**	**Score**
Fair	97	Good	92
Critical	62	Medium	62
Inappropriate	30	Unfair	29

3.3.8. Traffic and Pavement Index

Two other affecting parameters in bridge serviceability are traffic and pavement condition. For considering these parameters, the authors designed Table A10 in the questionnaire. Traffic effects are determined according to traffic volume and traffic condition. Pavement effects are considered as the pavement condition index (PCI). Therefore, the traffic and pavement index has three sub-indexes. Experts scored in the blank cells of Table A10 with the numbers from the range of 0 to 100. Furthermore, PCI is a number between 0 and 100. Table 17 shows the average of experts' views. Accordingly, the inspector needs three scores for each bridge: Traffic conditions score, Traffic volume score, and PCI. Finally, by averaging these three numbers, the score of traffic and pavement index will be obtained. It is important to point out that all sub-indexes (traffic condition, traffic volume, and pavement condition index) have the same importance.

Table 17. Scores of traffic sub-index.

Traffic Sun-Indexes Scores			
Traffic Conditions	**Score**	**Traffic Volume**	**Score**
Very good (traffic facilities are perfectly working, full sight distance and the number of lanes is standard)	95	Low	89
Good (traffic facilities are in relatively good condition, sight distance is desirable in most areas and the number of lanes is appropriate)	74	medium	68
Moderate (Some of traffic facilities are in bad conditions and the bridge has an undesirable curve)	51	Heavy	51
Bad (lanes are not enough, traffic facilities are not working, the bridge has a horizontal and vertical curve together, the sight distance is not appropriate).	12	Very heavy	26

4. Results and Discussion

The theoretical method needs to test in the real world to be more reliable for engineers. For this purpose, in this study, five bridges in Semnan province in Iran have been selected and implemented the proposed method for them. These bridges as following:

- Bridge No. 1: The bridge of Shahmirzad road intersection,
- Bridge No. 2: The bridge of Sari road intersection,
- Bridge No. 3: The bridge on 73rd km of Semnan-Damghan road,
- Bridge No. 4: The bridge on 6th km of Semnan-Jandaq road,
- Bridge No. 5: The bridge on 12th km of Semnan-Jandaq road.

First, all bridges are inspected by the authors. After that, the BCI of all bridges was determined based on the proposed method in this study. Finally, the prioritization of all bridges was conducted. The results of inspection and rating of bridges are given in the following subsections.

4.1. Determination of BCI in Bridge No. 1

The bridge of Shahmirzad road intersection is located in the city of Semnan, at the beginning of the Semnan-Shahmirzad road (Figure 5). This bridge has two spans and acts as the overpass of the Mashhad-Tehran highway. The bridge is forty-five meters in length, twenty-three meters in width, and has three lanes in each direction.

As shown in Figures 6 and 7, the structural elements of this bridge were in satisfactory conditions. Mostly, these elements were in low-distress conditions and rarely had medium-distress conditions. Hydrology and climate condition was good. The safety of the bridge was moderate. This bridge is part of the highway network, and the load impact score was forty-five. Earth type is II, and the seismic area type is high relative risky. The strategic importance of the bridge is medium. Bridge facilities were in weak condition. The drainage system and the electrical facilities were in inappropriate and medium conditions, respectively. Some of the traffic facilities were in bad condition, traffic volume was medium, and PCI was 76 (Figure 8). After completing the inspection and determination of all indices scores, BCI can be calculated (Table 18).

Figure 5. The bridge No. 1 on Google Earth (Latitude: 35°36′15″ N, Longitude: 53°22′16″ E).

Figure 6. The structure of the bridge No. 1.

Figure 7. An example of minor cracking in the wall of the bridge No. 1.

Figure 8. The pavement condition in the bridge No. 1.

Table 18. Calculation of BCI in bridge No. 1.

	Index	W_i	X_i	$W_i \times X_i$	BCI $= \sum (W_i \times X_i)$
1	Structural	0.331	93.765	31.036	
2	Hydrology and Climate	0.097	95.5	9.264	
3	Safety	0.146	66	9.636	
4	Bridge Performance (load impact)	0.08	45	3.6	72.849
5	Geotechnical and Seismic	0.143	55.5	7.937	
6	Strategic Importance	0.088	55	4.84	
7	Facilities	0.046	46	2.116	
8	Traffic and Pavement	0.068	65	4.42	

4.2. Determination of BCI in Bridge No. 2

The bridge of Sari road intersection is located in Damghan, at the cross of Damghan-Semnan and Damghan-Sari roads (Figure 9). The bridge has two spans and two lanes and is twenty-five meters in length and nine meters in width. After inspecting, the structural elements of this bridge, including deck, girder, bent-abutment-wall, and foundation, received the highest possible score because of their health conditions (Figure 10). In hydrology and climate examinations, there was no problem. The only problem of safety was related to lighting and brightness. The scores of the load impact, geotechnical and seismic, and strategic importance indices were similar to bridge No. 1. Another issue was related to the mechanical facilities of the drainage system that was inappropriate (Figure 11). Traffic condition was very well, volume traffic was medium, and PCI was 100. A summary of the BCI calculations in this bridge is presented in Table 19.

Figure 9. The bridge No. 2 on Google Earth (Latitude: 36°10′35″ N, Longitude: 54°18′17″ E).

Figure 10. The structural elements of bridge No. 2.

Figure 11. An example of block in the facilities of drainage system (bridge No. 2).

Table 19. BCI calculation of the bridge No. 2.

	Index	W_i	X_i	$W_i \times X_i$	BCI = $\sum (W_i \times X_i)$
1	Structural	0.331	93.765	31.036	
2	Hydrology and Climate	0.097	95.5	9.264	
3	Safety	0.146	63	9.198	
4	Bridge Performance (load impact)	0.08	45	3.6	73.221
5	Geotechnical and Seismic	0.143	55.5	7.937	
6	Strategic Importance	0.088	55	4.84	
7	Facilities	0.046	30	1.38	
8	Traffic and Pavement	0.068	87.67	5.962	

4.3. Determination of BCI in Bridge No. 3

This bridge is located at 73rd km of Semnan-Damghan road (Figure 12). The bridge has five spans, a total length of sixty meters, a width of ten meters, and two lanes. Similar to the two previous bridges, this bridge had low damage in its structural elements. Figures 13 and 14 show the structure and the example of the structural damage in this bridge, respectively. In hydrology and climate index, status was moderate. The most important issue was about the quality of protection against destructive matters. Lighting and brightness conditions were not proper and can cause safety issues in the bridge. This bridge is part of the highway network, and therefore the score of the load impact index was forty-five. Earth type is II, and the bridge is located in a high relative risk region. The strategic importance of the bridge is medium. The drainage system was another issue in this bridge because of its critical condition. Traffic facilities acted perfectly, sight distance was fully covered, and the number of lanes was standard. The traffic volume was medium, and the pavement was in very good condition with some minor distresses in the shoulder (Figure 15). A summary of the BCI calculation of this bridge is presented in Table 20.

Figure 12. The bridge No. 3 on Google Earth (Latitude: 35°57′22″ N, Longitude: 54°01′54″ E).

Figure 13. The structure of bridge No. 3.

Figure 14. An example of structural damage in bridge No. 3.

Figure 15. The pavement condition of bridge No. 3.

Table 20. BCI calculation of the bridge No. 3.

	Index	W_i	X_i	$W_i \times X_i$	BCI = $\sum (W_i \times X_i)$
1	Structural	0.331	93.765	31.036	
2	Hydrology and Climate	0.097	66	6.402	
3	Safety	0.146	72.33	10.56	
4	Bridge Performance (load impact)	0.08	45	3.6	73.193
5	Geotechnical and Seismic	0.143	55.5	7.937	
6	Strategic Importance	0.088	55	4.84	
7	Facilities	0.046	62	2.852	
8	Traffic and Pavement	0.068	87.67	5.962	

4.4. Determination of BCi in Bridge No. 4

This bridge is located at 6th km of Semnan-Jandaq road (Figure 16). It has one span, a length of eight meters, a width of seven meters, and two lanes. All structural subindexes were in low damage condition, except the bent-abutment-wall that had mediate damage intensity (Figure 17). Although this bridge is exposed to invasive agents (look at Figure 16), there was not appropriate protection against this issue. For this reason, the score of hydrology and climate index decreased into the moderate range. The safety of the bridge was not in a satisfactory status. Safety equipment needed repair, lighting condition is critical, and drainage condition was not very well (look at Figure 18). This bridge services as a member of a minor road, and therefore, the score of the load impact index is seventy. Earth type is III, and the bridge is located in a region with high relative risk. Based on Table 15, the score of strategic importance is twenty-nine because this bridge does not link the strategic areas, streets, and non-strategic arterials. In this bridge, there were no electrical facilities, and mechanical facilities were not in appropriate condition. Therefore, the score of the facilities index was thirty. Traffic volume was very low, traffic condition was very bad, and also, the pavement had not satisfactory status, and PCI was thirty-four (Figure 19). Table 21 shows a summary of the BCI calculation of this bridge.

Figure 16. The bridge No. 4 on Google Earth (Latitude: 35°32′28″ N, Longitude: 53°29′49″ E).

Figure 17. The structural damage in bridge NO. 4.

Figure 18. Safety and drainage issues in bridge No. 4.

Figure 19. Condition of pavement distresses in bridge NO. 4.

Table 21. BCI calculation of the bridge on 6th km of Semnan-Jandaq road.

	Index	W_i	X_i	$W_i \times X_i$	$BCI = \sum (W_i \times X_i)$
1	Structural	0.331	86.355	28.583	
2	Hydrology and Climate	0.097	62.75	6.087	
3	Safety	0.146	52.67	7.69	
4	Bridge Performance (load impact)	0.08	70	5.6	
5	Geotechnical and Seismic	0.143	43.5	6.22	62.172
6	Strategic Importance	0.088	29	2.552	
7	Facilities	0.046	30	1.38	
8	Traffic and Pavement	0.068	45	3.06	

4.5. Determination of BCI in Bridge No. 5

This bridge is located at 12th km of Semnan-Jandaq road (Figure 20) and has three spans, a length of twenty meters, a width of eight meters, and two lanes. Due to problems in the wall and foundation (Figures 21 and 22), the score of the structural index decreases to 83.635. Invasive agents caused erosion, and protection against them is weak. Safety equipment required major repair, lighting and brightness condition was in a bad status, and drainage condition could not provide the desired friction (look at Figure 23). Because of being in the same region and transport network, bridges No.4 and No. 5 have similar scores in the load impact index, geotechnical and seismic index, and strategic importance index. The mechanical facilities of the drainage system were improper. The volume of traffic was at a level of low. There are no proper traffic horizontal/vertical signs. The pavement had various distresses, and PCI was thirty-six. BCI in this bridge is calculated in Table 22.

Table 22. BCI calculation of the bridge on 12th km of Semnan-Jandaq road.

	Index	W_i	X_i	$W_i \times X_i$	$BCI = \sum (W_i \times X_i)$
1	Structural	0.331	83.635	27.683	
2	Hydrology and Climate	0.097	52.75	5.117	
3	Safety	0.146	35	5.11	
4	Bridge Performance (load impact)	0.08	70	5.6	
5	Geotechnical and Seismic	0.143	43.5	6.22	56.768
6	Strategic Importance	0.088	29	2.552	
7	Facilities	0.046	30	1.38	
8	Traffic and Pavement	0.068	45.67	3.106	

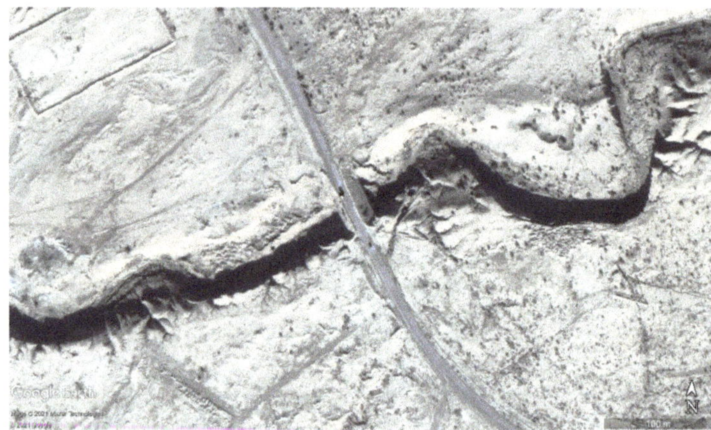

Figure 20. The bridge No. 5 on Google Earth (Latitude: 35°31′59″ N, Longitude: 53°30′46″ E).

Figure 21. Damage in the wall of the bridge No. 5.

Figure 22. Bridge scour in bridge No. 5.

Figure 23. Distresses of pavement and safety problems in bridge No. 5.

In this section, the authors aimed to check the ability of their methodology. They selected five case studies from the bridge network in Semnan province, in Iran. The authors tried to adopt the bridges that have the maximum possible difference in condition. According to Tables 18–22, the bridge of 12th km of Semnan-Jandaq road has the lowest BCI among the five studied bridges (BCI = 56.8). This bridge takes the highest priority for maintenance. The second priority is related to the bridge of 6th km of Semnan-Jandaq road with BCI = 62.2. The bridge at the beginning of Shahmirzad road, with BCI = 72.8, takes the third priority. Two other bridges with BCI = 73.2 take the lowest maintenance priority.

As mentioned in the earlier sections, a review of previous studies shows the lack of a comprehensive method for evaluating and prioritizing bridges. Each of the methods proposed by other researchers generally focuses on limited parts of the factors affecting the bridge conditions. In this study, the authors tried to develop a new methodology that includes all the factors affecting the condition of bridges. Scrutiny of the results obtained from five under-study bridges, with different characteristics, confirms that this method is feasible. On the one hand, the method is simple and can save time and money in the health monitoring process of the bridge network. On the other hand, the flexibility of the methodology is high, and therefore this methodology can be easily calibrated and implemented in any other place. It is enough to design the relevant questionnaire, gather the opinion of bridge experts, analyze the filled questionnaires, determine scores and weights, and after that, inspect and prioritize bridges. This study helps remove another gap: The lack of an efficient, comprehensive method/system for health monitoring of bridges in Iran. This gap results in wasting resources and time. Continuous, exact, complete health monitoring and correctly prioritizing are essential needs for infrastructures. The proposed methodology can simply apply in various regions of Iran. Therefore, this study helps improve the quality of BMS activities in Iran, and it can be another helpfulness of this work.

Despite the positive points mentioned in the previous paragraph, the proposed method has a limitation named the subjective influence of inspectors. Visual inspection is a measurement mechanism implemented by humans. Accordingly, variability influences the reliability of this mechanism. In fact, the visual examination gives valuable data on bridge health, but it is not always guaranteed since it depends mainly on the inspector's experience and knowledge. Of course, this issue can hardly be avoided. One solution for such an issue is to use auxiliary analyses, such as non-destructive tests (NDTs). Although this solution can confine the subjective influence of inspectors, it will confront the authors with another limitation. Most of the local organizations in Iran are deprived of NDT equipment or similar tools. Moreover, the authors intend to present an easy-to-use, applicable method for all organizations in Iran, including local organizations. Eventually, the authors decided to base their methodology according to direct inspection by inspectors. In their opinion, the advantages of being more usable of the methods conquer possible disadvantages of subjective influence of inspectors. For solving or reducing the problem of inspectors' influence, they suggest that organizations should focus on the personal selection and better training of inspectors.

As part of future works, the authors are investigating three different research objectives. First, the authors are very interested in connecting new technologies/tools/methods with their methodology. One of these new technologies is remote sensing (RS). RS analyzes different objects on the earth's surface by data received from a device that is not in contact with those objects. Another tool is machine learning (ML). ML originated from artificial intelligence (AI) and has been used in recent years in various scientific areas. Increasing applications of RS [48–54] and ML [55–59] in structural health monitoring of infrastructures in recent years motivates the authors to conduct this idea. Second, the authors intend to extend the proposed methodology for other types of bridges, including steel and stone (old) bridges. The latter idea can help enlarge the application dominance of the methodology. Finally, by increasing the number of experts asked to fill the questionnaire, the accuracy and efficiency of the methodology will be more reliable.

5. Conclusions

This study presents a new methodology for the determination of BCI in concrete bridges. BCI constitutes eight indices and several sub-indexes. Each one of these indices and sub-indices has a specific score and importance weight. The scores and weights are assigned by experts of bridge engineering. After determining scores and weights, inspectors survey the bridge and assign the scores to all sub-index based on their condition. Then, the score of each index is obtained. Finally, by summing the weighted scores of indices, BCI will be determined. The necessity of this research could be justified in the absence of any comprehensive and effective system or index for assessing the bridge conditions, especially in Iran. Due to financial constraints and the lack of qualified specialists, it is also crucial to provide solutions to overcome these shortcomings. Therefore, in this research, attempts were made to develop a new, simple method for assessing bridge conditions in order to optimize the management activities. The novelty of this study is in the scoring system because the scoring system is constructed by native experts' views. On the other hand, because of the lack of a comprehensive, proper index in Iran, this paper is considered innovative. Simplicity is one of the characteristics of the proposed method because it does not require the application of non-destructive equipment or laboratory tests. The method allows measuring BCI and prioritizing bridges for maintenance based on the visual evaluation of the damages and general characteristics of the bridge and their performance. Therefore, time and budget can be saved in this method. On the other hand, the experts' views, scores, and coefficients of relative importance may vary in different organizations or countries. Thus, the calibration of this method is only done by designing the questionnaire and collecting experts' views. For testing the proposed method, five bridges in Semnan province were inspected, and their BCI was determined to prioritize bridges.

Author Contributions: Conceptualization, S.D. and H.G.T.; methodology, H.G.T. and N.K.; software, S.D. and N.K.; validation, H.G.T., N.K., and A.M.; formal analysis, S.D. and H.G.T.; investigation, N.K. and A.M.; resources, S.D.; data curation, H.G.T. and N.K.; writing—original draft preparation, S.D. and H.G.T.; writing—review and editing, N.K.; visualization, A.M.; supervision, H.G.T. and A.M.; project administration, H.G.T.; funding acquisition, A.M. All authors have read and agreed to the published version of the manuscript.

Funding: This research in part is supported by the project GINOP-2.2.1-18-2018-00015.

Data Availability Statement: Not applicable.

Conflicts of Interest: The authors declare no conflict of interest.

Appendix A. Survey Questionnaire

This appendix presents the questionnaire used in this study.

Appendix A.1. Introduction

This survey belongs to a study that will provide a bridge condition index (BCI) for concrete bridges in Iran. BCI is an applicable tool that can be help engineers in bridge management programs. This tool provides a method for evaluating and prioritizing the existing bridges. Because this method will be implemented for bridge networks in Iran, the viewpoint of Iranian experts is crucial in this study. Therefore, this questionnaire aims to gather the opinion of bridge experts about influencing indexes on bridge conditions in Iran.

Appendix A.2. Overview

BCI involves eight indices, including Structural, Hydrology and Climate, Safety, Bridge Performance, Geotechnical and Seismic, Strategic Importance, Facilities, and Traffic and Pavement. Moreover, each index divides into several sub-indexes. Therefore, you deal with these indexes and sub-indexes.

This survey consists of several tables, which you must insert your standpoint as a number. The necessary explanations are presented in each part and help you.

Appendix A.2.1. BCI Indices

Table A1 aims to provide the relative importance of affecting indices on bridge conditions. You must assign a value between 0 and 10 based on your technical experience and expertise in each cell of the table. If the row index is more important than the column index, a value between 1 and 10 must be assigned (the more important, the bigger the number). In contrast, if the column index is more important than the row index, a value between 0 and 1 must be assigned (the more important, the smaller the number). In this pairwise comparison, the number one is related to the same relative importance.

Table A1. Relative importance of indices.

Index	Structural	Hydrology and Climate	Safety	Bridge Performance (Load Impact)	Geotechnical and Seismic	Strategic Importance	Facilities	Traffic and Pavement
Structural	1							
Hydrology and Climate		1						
Safety			1					
Bridge Performance (load impact)				1				
Geotechnical and Seismic					1			
Strategic Importance						1		
Facilities							1	
Traffic and Pavement								1

Appendix A.2.2. Structural Index

The structural index includes four sub-indexes. Table A2 aims to determine the relative importance of these sub-indexes. Please, assign a value between 0 and 10 based on your technical experience and expertise in each cell of the table. If the row index is more important than the column index, a value between 1 and 10 must be assigned (the more important, the bigger the number). In contrast, if the column index is more important than the row index, a value between 0 and 1 must be assigned (the more important, the smaller the number). In this pairwise comparison, the number one is related to the same relative importance.

Table A2. Relative Importance of Sub-Indices in Structural index.

Sub-Index	Deck	Girder	Bent-Abutment-Wall	Foundation
Deck	1			
Girder		1		
Bent-abutment-wall			1	
Foundation				1

Based on the damage intensity, score four sub-indexes of the structural index in Table A3. Your scores must be in the range of 0 to 100. Note that the score of 100 is related to the best condition and the score of 0 is related to the worst condition.

Table A3. Scores of Sub-Indexes in Structural Index.

Damage Intensity	Sub-Indexes Scores			
	Deck	Girder	Bent-Abutment-Wall	Foundation
Low				
Mediate				
High				

Appendix A.2.3. Hydrology and Climate Index

Based on the explanation of Table A4, rate river condition, river type, climatic feature, and the existence of destructive agent. Your scores must be in the range of 0 to 100. Note that the score of 100 is related to the best condition and the score of 0 is related to the worst condition. It is important to point out that all sub-indexes have the same importance. Therefore, the overall score of the index includes the sum of sub-indexes scores with equal weights.

Appendix A.2.4. Safety Index

Based on the explanation of Table A5, provide the proper score for various conditions of safety equipment. Your scores must be in the range of 0 to 100. Note that the score of 100 is related to the best condition and the score of 0 is related to the worst condition. It is important to point out that all sub-indexes have the same importance. Therefore, the overall score of the index includes the sum of sub-indexes scores with equal weights.

Table A4. Scores of sub-indexes in hydrology and climate index.

Sub-Indexes Scores			
River Condition		**River Type**	
Description	**Score**	**Type**	**Score**
There is no erosion in the riverbed or the erosion is trivial. The amount of sedimentation and debris is negligible		Area under the bridge is not a river path	
The riverbed has eroded slightly. There are signs of depositions in the upstream and downstream. Further analysis is required to detect failures		There is seasonal river flowing under the bridge.	
The erosion of the riverbed is critical and concerning. There are enormous amounts of sedimentations around the bridge. Serious measures have to be taken.		There is permanent river flowing under the bridge.	
Climatic Features		**Destructive Agents**	
Description	**Score**	**Quality of Protection against Destructive Matters**	**Score**
Mild (there are no invasive agents such as moisture, transpiration, freezing and melting cycle, corrosive substances, etc.)		Very good	
Medium (conditions that are occasionally exposed to moisture and transpiration, and elements that are permanently exposed to non-invasive soils and water, or underwater with a pH > 5)		Good	
Severe (extreme humidity or transpiration, or freezing and thawing cycle, elements immersed in water, such that one surface is exposed to air, elements in chlorine ion air, elements exposed to corrosion caused by the use of anti-freezing agents)		Medium	
Extremely severe (conditions that are exposed to gases, water and static sewage with a pH of up to 5, corrosive matters, moisture with extreme icing and melting)		Bad	
Exceptionally severe (conditions subject to extreme erosion, flowing water and sewage with a maximum pH of 5)			

Table A5. Scores of sub-indexes in Safety index.

Sub-Indexes Scores					
Curbs, Guardrails and Fences		**Lighting and Brightness**		**Drainage of Surface Water**	
Description of Defects	**Score**	**Conditions**	**Score**	**Drainage Condition**	**Score**
No repair is needed		Trivial dazzling, excellent color rendering, broad sight		Perfect drainage, adequate friction coefficient	
Partial repair is needed		Slight dazzling, color rendering and sight are relatively desirable		Drainage for securing desirable friction	
Major repair is required		Extreme dazzling, low color rendering and limited sight		Improper drainage, undesirable friction coefficient	

Appendix A.2.5. Load Impact Index

Based on the class and the type of transport (road or rail), provide the proper score for Table A6. Your scores must be in the range of 0 to 100.

Table A6. Load impact score.

Class	Transport Type	
	Road	Rail
Freeway		
Highway and major road		
Minor road		
Rural road		
Metro and monorail		

Appendix A.2.6. Geotechnical and Seismic Index

Based on the earth type and the seismic area type, give the appropriate score for Table A7. Your scores must be in the range of 0 to 100. Note that the score of 100 is related to the best and the safest situation, and the score of 0 is related to the worst and most risky one. It is important to point out that both sub-indexes have the same importance. Therefore, the overall score of the index includes the sum of sub-indexes scores with equal weights.

Table A7. Scores of Sub-Indexes in Geotechnical and Seismic Index.

Sub-Indexes Scores			
Geotechnical		Seismic	
Earth Type	Score	Seismic Area Type	Score
I		Low relative risk	
II		Medium relative risk	
III		High relative risk	
IV		Very high relative risk	

Appendix A.2.7. Strategic Importance Index

Table A8 evaluates the strategic importance of bridges. Write the appropriate scores. Your scores must be in the range of 0 to 100.

Table A8. Scores of the strategic importance index.

The Strategic Importance of Bridge	Score
High importance (links two strategic areas)	
Medium importance (links streets and non-strategic arterial)	
Low importance (other bridges)	

Appendix A.2.8. Facilities Index

Considering conditions of mechanical and electrical facilities can be conducted in Table A9. Please, write the appropriate scores in the range of 0 to 100. Note that the score of 100 is related to the best condition, and the score of 0 is related to the worst condition. It is important to point out that both sub-indexes of Table A9 have the same importance. Therefore, the overall score of the facilities index concludes from averaging of sub-indexes scores with equal weights.

Table A9. Scores of sub-indexes in facilities index.

Sub-Indexes Scores			
Mechanical Facilities		Electrical Facilities	
Drainage System	Score	Lighting Condition	Score
Fair		Good	
Critical		Medium	
Inappropriate		Unfair	

Appendix A.2.9. Traffic and Pavement Index

This index is related to traffic and pavement conditions in bridges. It has three sub-indexes, including traffic condition, traffic volume, and the pavement condition index. Fill appropriate scores in the blank cells of Table A10. This table represents the traffic condition and volume of a bridge. Your scores must be in the range of 0 to 100. Whatever the traffic condition is better and traffic volume is lower, your score must be higher because the bridge is in a more reliable situation. Furthermore, for considering the pavement condition, we use the pavement condition index (PCI). PCI is an index that ranges from 0 to 100. It is important to point out that all sub-indexes (traffic condition, traffic volume, and pavement condition index) have the same importance. Therefore, the overall score of the index includes the sum of sub-indexes scores with equal weights.

Table A10. Scores of traffic sub-index.

Traffic Sub-Indexes Scores			
Traffic Conditions	Score	Traffic Volume	Score
Very good (traffic facilities are perfectly working, full sight distance and the number of lanes is standard)		Low	
Good (traffic facilities are in relatively good condition, sight distance is desirable in most areas and the number of lanes is appropriate)		medium	
Moderate (Some of traffic facilities are in bad conditions and the bridge has an undesirable curve)		Heavy	
Bad (lanes are not enough, traffic facilities are not working, the bridge has a horizontal and vertical curve together, the sight distance is not appropriate).		Very heavy	

Appendix A.3. Responding Information

Here is some information that can help us. Please, provide them:
Your name (optional):
Agency/University/Company:
Current position:
Address (optional):
Date:
E-mail:
Please, submit your completed questionnaire to: N.Karballaeezadeh@shahroodut.ac.ir or N.karballaeezadeh@gmail.com
Thank you in advance for your support and cooperation with our study.

References

1. Liu, H.; Wang, X.; Tan, G.; He, X. System Reliability Evaluation of a Bridge Structure Based on Multivariate Copulas and the AHP–EW Method That Considers Multiple Failure Criteria. *Appl. Sci.* **2020**, *10*, 1399. [CrossRef]
2. Jeong, Y.; Kim, W.; Lee, I.; Lee, J. Bridge inspection practices and bridge management programs in China, Japan, Korea, and US. *J. Struct. Integr. Maint.* **2018**, *3*, 126–135.

3. Patel, D.; Lad, V.; Chauhan, K.; Patel, K. Development of Bridge Resilience Index Using Multicriteria Decision-Making Techniques. *J. Bridge Eng.* **2020**, *25*, 04020090. [CrossRef]
4. Liu, H.; Wang, X.; Tan, G.; He, X.; Luo, G. System reliability evaluation of prefabricated RC hollow slab Bridges considering hinge joint damage based on modified AHP. *Appl. Sci.* **2019**, *9*, 4841. [CrossRef]
5. Mohammadzadeh, D.; Karballaeezadeh, N.; Zahed, A.S.; Mosavi, A.; Imre, F. Three-Dimensional Modeling and Analysis of Mechanized Excavation for Tunnel Boring Machines. *Acta Polytech. Hung.* **2021**, *18*, 213–230. [CrossRef]
6. Akgul, F. Inspection and evaluation of a network of concrete bridges based on multiple NDT techniques. *Struct. Infrastruct. Eng.* **2020**, 1–20. [CrossRef]
7. Ryall, M. *Bridge Management*; CRC Press: Boca Raton, FL, USA, 2009.
8. Wu, C.; Wu, P.; Wang, J.; Jiang, R.; Chen, M.; Wang, X. Critical review of data-driven decision-making in bridge operation and maintenance. *Struct. Infrastruct. Eng.* **2020**, 1–24. [CrossRef]
9. Zhang, L.; Qiu, G.; Chen, Z. Structural health monitoring methods of cables in cable-stayed bridge: A review. *Measurement* **2021**, *168*, 108343. [CrossRef]
10. Jeong, E.; Seo, J.; Wacker, J. Literature Review and Technical Survey on Bridge Inspection Using Unmanned Aerial Vehicles. *J. Perform. Constr. Facil.* **2020**, *34*, 04020113. [CrossRef]
11. Nili, M.H.; Taghaddos, H.; Zahraie, B. Integrating discrete event simulation and genetic algorithm optimization for bridge maintenance planning. *Autom. Constr.* **2021**, *122*, 103513. [CrossRef]
12. Moodi, F. Development of a Knowledge Based Expert System for the Repair and Maintenance of Concrete Structures. Ph D Thesis, Newcastle upon Tyne University, Newcastle upon Tyne, UK, 2001.
13. Rashidi, M.; Gibson, P. A methodology for bridge condition evaluation. *J. Civ. Eng. Archit.* **2012**, *6*, 1149–1157.
14. Wu, H.-C. *A Multi-Objective Decision Support Model for Maintenance and Repair Strategies in Bridge Networks*; Columbia University: New York, NY, USA, 2008.
15. Abu Dabous, S.; Alkass, S. Decision support method for multi-criteria selection of bridge rehabilitation strategy. *Constr. Manag. Econ.* **2008**, *26*, 883–893. [CrossRef]
16. Wakchaure, S.S.; Jha, K.N. Determination of bridge health index using analytical hierarchy process. *Constr. Manag. Econ.* **2012**, *30*, 133–149. [CrossRef]
17. The Standardization Administration of the People's Republic of China. *Standards for Technical Condition Evaluation of Highway Bridges-JTG/T H21-2011*; The Standardization Administration of the People's Republic of China: Beijing, China, 2011.
18. MHURD. *Technical Code of Maintenance for City Bridge (CJJ99-2003)*; Ministry of Housing and Urban-Rural Development: Beijing, China, 2003.
19. MLIT. *Manual for Bridge Periodic Inspection*; Ministry of Land, Infrastructure, Transport, and Tourism: Tokyo, Japan, 2014.
20. MOLIT. *Guideline of Safety Inspection and In-Depth Safety Inspection for Structures*; Ministry of Land, Infrastructure and Trasport: Seoul, Korea, 2012.
21. Shepard, R.W.; Johnson, M.B. *California Bridge Health Index: A Diagnostic Tool to Maximize Bridge Longevity, Investment*; Transportation Research Board: Washington, DC, USA, 2001.
22. FHWA. *Bridge Inspector's Reference Manual-FHWA NHI 12-049*; Federal Highway Administration: Arlington, VA, USA, 2012.
23. Rashidi, M.; Gibson, P.; Ho, T.K. *A New Approach to Bridge Infrastructure Management*; International Symposium for Next Generation Infrastructure: Wollongong, Australia, 2013.
24. Rashidi, M.; Samali, B.; Sharafi, P. A new model for bridge management: Part A: Condition assessment and priority ranking of bridges. *Aust. J. Civ. Eng.* **2016**, *14*, 35–45. [CrossRef]
25. Rashidi, M.; Samali, B.; Sharafi, P. A new model for bridge management: Part B: Decision support system for remediation planning. *Aust. J. Civ. Eng.* **2016**, *14*, 46–53. [CrossRef]
26. Rashidi, M.; Lemass, B.P. A decision support methodology for remediation planning of concrete bridges. *KICEM J. Constr. Eng. Proj. Manag.* **2011**, *1*, 2. [CrossRef]
27. Akgul, F. Bridge management in Turkey: A BMS design with customised functionalities. *Struct. Infrastruct. Eng.* **2016**, *12*, 647–666. [CrossRef]
28. Karimzadeh, S.; Matsuoka, M. Remote Sensing X-Band SAR Data for land subsidence and pavement monitoring. *Sensors* **2020**, *20*, 4751. [CrossRef]
29. Karballaeezadeh, N.; Mohammadzadeh, S.D.; Moazemi, D.S.; Band, S.; Mosavi, A.; Reuter, U. Smart Structural Health Monitoring of Flexible Pavements Using Machine Learning Methods. *Coatings* **2020**, *10*, 1100. [CrossRef]
30. Bitarafan, M.; Zolfani, S.H.; Arefi, S.L.; Zavadskas, E.K.; Mahmoudzadeh, A. Evaluation of real-time intelligent sensors for structural health monitoring of bridges based on SWARA-WASPAS; a case in Iran. *Balt. J. Road Bridge Eng.* **2014**, *9*, 333–340. [CrossRef]
31. Yousefi, Y.; Karballaeezadeh, N.; Moazami, D.; Zahed, A.S.; Mosavi, A. Improving aviation safety through modeling accident risk assessment of runway. *Int. J. Environ. Res. Public Health* **2020**, *17*, 6085. [CrossRef]
32. Lima, M.M.; Limaei, S.A.A. Structural health monitoring of concrete bridges in Rudbar-Manjil region in Iran. In *Proceedings of the International Conference on Intelligent Building and Management*; IACSIT Press: Singapore, 2011; pp. 323–327.
33. Abdollahzadeh Nasiri, A.S.; Rahmani, O.; Abdi Kordani, A.; Karballaeezadeh, N.; Mosavi, A. Evaluation of Safety in Horizontal Curves of Roads Using a Multi-Body Dynamic Simulation Process. *Int. J. Environ. Res. Public Health* **2020**, *17*, 5975. [CrossRef]

34. Emadali, L.; Motagh, M.; Haghighi, M.H. Characterizing post-construction settlement of the Masjed-Soleyman embankment dam, Southwest Iran, using TerraSAR-X SpotLight radar imagery. *Eng. Struct.* **2017**, *143*, 261–273. [CrossRef]

35. Nabipour, N.; Karballaeezadeh, N.; Dineva, A.; Mosavi, A.; Mohammadzadeh, S.D.; Shamshirband, S. Comparative analysis of machine learning models for prediction of remaining service life of flexible pavement. *Mathematics* **2019**, *7*, 1198. [CrossRef]

36. Jahan, S.; Mojtahedi, A.; Mohammadyzadeh, S.; Hokmabady, H. A Fuzzy Krill Herd Approach for Structural Health Monitoring of Bridges using Operational Modal Analysis. *Iran. J. Sci. Technol. Trans. Civ. Eng.* **2020**, *45*, 1139–1157. [CrossRef]

37. Karballaeezadeh, N.; Ghasemzadeh Tehrani, H.; Mohammadzadeh Shadmehri, D.; Shamshirband, S. Estimation of flexible pavement structural capacity using machine learning techniques. *Front. Struct. Civ. Eng.* **2020**, *14*, 1083–1096. [CrossRef]

38. Rogulj, K.; Kilić Pamuković, J.; Jajac, N. Knowledge-Based Fuzzy Expert System to the Condition Assessment of Historic Road Bridges. *Appl. Sci.* **2021**, *11*, 1021. [CrossRef]

39. Agrawal, A.K.; Kawaguchi, A.; Chen, Z. Deterioration rates of typical bridge elements in New York. *J. Bridge Eng.* **2010**, *15*, 419–429. [CrossRef]

40. Hsu, H.; Chang, W.; Wang, R.; Cho, C.; Jiang, D. Small and medium size bridge maintenance sequence analysis by optimization technique. In *Advances in Bridge Maintenance, Safety Management, and Life-Cycle Performance, Set of Book & CD-ROM*; CRC Press: Boca Raton, FL, USA, 2015; pp. 139–140.

41. Hearn, G. Condition data and bridge management systems. *Struct. Eng. Int.* **1998**, *8*, 221–225. [CrossRef]

42. Ikpong, A.; Chandra, A.; Bagchi, A. Alternative to AHP Approach to Criteria Weight Estimation in Highway Bridge Management. *Can. J. Civ. Eng.* **2020**. [CrossRef]

43. Ramadhan, R.H.; Wahhab, H.I.A.A.; Duffuaa, S.O. The use of an analytical hierarchy process in pavement maintenance priority ranking. *J. Qual. Maint. Eng.* **1999**, *5*, 25–39. [CrossRef]

44. Prakasan, A.C.; Tiwari, D.; Shah, Y.U.; Parida, M. Pavement Maintenance Prioritization of Urban Roads Using Analytical Hierarchy Process. *Int. J. Pavement Res. Technol.* **2015**, *8*, 112–122.

45. Phung, X.L.; Truong, H.S.; Bui, N.T. Expert system based on integrated fuzzy AHP for automatic cutting tool selection. *Appl. Sci.* **2019**, *9*, 4308. [CrossRef]

46. Chen, L.; Deng, X. A modified method for evaluating sustainable transport solutions based on AHP and Dempster–Shafer evidence theory. *Appl. Sci.* **2018**, *8*, 563. [CrossRef]

47. Kuzman, M.K.; Grošelj, P.; Ayrilmis, N.; Zbašnik-Senegačnik, M. Comparison of passive house construction types using analytic hierarchy process. *Energy Build.* **2013**, *64*, 258–263. [CrossRef]

48. Milillo, P.; Giardina, G.; Perissin, D.; Milillo, G.; Coletta, A.; Terranova, C. Pre-collapse space geodetic observations of critical infrastructure: The Morandi Bridge, Genoa, Italy. *Remote Sens.* **2019**, *11*, 1403. [CrossRef]

49. Alamdari, M.M.; Ge, L.; Kildashti, K.; Zhou, Y.; Harvey, B.; Du, Z. Non-contact structural health monitoring of a cable-stayed bridge: Case study. *Struct. Infrastruct. Eng.* **2019**, *15*, 1119–1136. [CrossRef]

50. Gagliardi, V.; Benedetto, A.; Ciampoli, L.B.; D'Amico, F.; Alani, A.M.; Tosti, F. Health monitoring approach for transport infrastructure and bridges by satellite remote sensing Persistent Scatterer Interferometry (PSI). In *Earth Resources and Environmental Remote Sensing/GIS Applications XI*; International Society for Optics and Photonics Location: Washington, DC, USA, 2020; p. 115340K.

51. Alani, A.M.; Tosti, F.; Ciampoli, L.B.; Gagliardi, V.; Benedetto, A. An integrated investigative approach in health monitoring of masonry arch bridges using GPR and InSAR technologies. *NDT E Int.* **2020**, *115*, 102288. [CrossRef]

52. Marchewka, A.; Ziółkowski, P.; Aguilar-Vidal, V. Framework for structural health monitoring of steel bridges by computer vision. *Sensors* **2020**, *20*, 700. [CrossRef]

53. Rodriguez Polania, D. Bridges Structural Health Monitoring (SHM) with Aid of Building Information Modeling (BIM) and Remote Sensing Technologies. Doctoral Dissertation, Politecnico di Torino, Turin, Italy, 2020.

54. Gordan, M.; Ismail, Z.; Ghaedi, K.; Ibrahim, Z.; Hashim, H.; Ghayeb, H.H.; Talebkhah, M. A brief overview and future perspective of unmanned aerial systems for in-service structural health monitoring. *Eng. Adv.* **2021**, *1*, 9–15. [CrossRef]

55. Anaissi, A.; Khoa, N.L.D.; Mustapha, S.; Alamdari, M.M.; Braytee, A.; Wang, Y.; Chen, F. Adaptive one-class support vector machine for damage detection in structural health monitoring. In *Pacific-Asia Conference on Knowledge Discovery and Data Mining*; Springer: Cham, Switzerland, 2017; pp. 42–57.

56. Khoa, N.L.D.; Alamdari, M.M.; Rakotoarivelo, T.; Anaissi, A.; Wang, Y. Structural health monitoring using machine learning techniques and domain knowledge based features. In *Human and Machine Learning*; Springer: Cham, Switzerland, 2018; pp. 409–435.

57. Bao, Y.; Li, H. Machine learning paradigm for structural health monitoring. *Struct. Health Monit.* **2020**. [CrossRef]

58. Xiao, H.; Wang, W.; Dong, L.; Ogai, H. A Novel Bridge Damage Diagnosis Algorithm Based on Deep Learning with Gray Relational Analysis for Intelligent Bridge Monitoring System. *IEEJ Trans. Electr. Electron. Eng.* **2021**, *16*, 730–742. [CrossRef]

59. Sarmadi, H.; Entezami, A.; Salar, M.; De Michele, C. Bridge health monitoring in environmental variability by new clustering and threshold estimation methods. *J. Civ. Struct. Health Monit.* **2021**, 1–16. [CrossRef]

Article

pH Measurement of Cement-Based Materials: The Effect of Particle Size

Poh Yee Loh [1], Payam Shafigh [1,*], Herda Yati Binti Katman [2,*], Zainah Ibrahim [3] and Sumra Yousuf [4]

[1] Centre for Building, Construction & Tropical Architecture (BuCTA), Faculty of Built Environment, Universiti Malaya, Kuala Lumpur 50603, Malaysia; lohpohyee@gmail.com

[2] Department of Civil Engineering, University Tenaga Nasional, Kajang 43000, Malaysia

[3] Department of Civil Engineering, Faculty of Engineering, Universiti Malaya, Kuala Lumpur 50603, Malaysia; zainah@um.edu.my

[4] Department of Building and Architectural Engineering, Faculty of Engineering & Technology, Bahauddin Zakariya University, Multan 60000, Pakistan; sumra.yousafrm@gmail.com

* Correspondence: pshafigh@gmail.com (P.S.); Herda@uniten.edu.my (H.Y.B.K.)

Abstract: Healthy reinforced concrete should be highly alkaline to safeguard the passive protective film for reinforcement of steel bars against corrosion. pH measurement is gaining importance in research of cement-based materials (CBMs), such as paste, mortar and concrete, as well as in structural health monitoring and forensic engineering applications. However, insufficient information is available regarding the most practical, economical and applicable quantitative pH measurement method for CBMs from the sampling to measurement stage. Existing recommended methods for measuring pH have many variables that need to be investigated to determine how they influence the pH value. Samples were recommended to be ground into very fine particles for pH measurement. Preparing very fine particles of CBMs is costly and time consuming, while larger particles, with sizes similar to sand particles, are easier to obtain, without needing special equipment. This study aims to investigate the effect of different particle sizes on the pH of cement mortar. Mortar specimens were crushed and sieved to obtain different ranges of particle sizes to measure the pH values. Results showed that specimens with large particle sizes (between 600 µm and 4.75 mm) can produce similar results to specimens with very fine particle sizes (<600 µm) by increasing the solid-to-solvent ratio or the leaching time.

Keywords: pH measurement; pH value; ex-situ leaching; particle size; cement-based materials

Citation: Loh, P.Y.; Shafigh, P.; Katman, H.Y.B.; Ibrahim, Z.; Yousuf, S. pH Measurement of Cement-Based Materials: The Effect of Particle Size. *Appl. Sci.* **2021**, *11*, 8000. https://doi.org/10.3390/app11178000

Academic Editors: Hugo Filipe Pinheiro Rodrigues and Ivan Duvnjak

Received: 16 June 2021
Accepted: 1 August 2021
Published: 29 August 2021

Publisher's Note: MDPI stays neutral with regard to jurisdictional claims in published maps and institutional affiliations.

Copyright: © 2021 by the authors. Licensee MDPI, Basel, Switzerland. This article is an open access article distributed under the terms and conditions of the Creative Commons Attribution (CC BY) license (https://creativecommons.org/licenses/by/4.0/).

1. Introduction

Calcium silicates (C_3S and C_2S) are the main cementitious compounds in cement. The hydration and hydrolysis of calcium silicates produces calcium hydroxide ($Ca(OH)_2$ or CH) [1]. About 25% of the structural component of cement paste is CH, which increases the pH of cement-based materials (CBMs) to more than 12 [2]. The high pH of concrete will enable the formation of a passive protective film on the embedded steel. This thin oxide layer formed on the steel can prevent the metal atoms from dissolving and reduce the corrosion rate to an insignificant level [3]. However, carbonation, CH leaching, or the presence of ion chloride reduces the amount of CH and other alkalis in concrete [4]. When the pH of concrete is reduced to less than 11.5, the passive protective film for the embedded steel will be unstable and the steel will be corroded at the rate of at least 1000 times higher [3–5]. Phenolphthalein solution is typically used as an indicator to test the depth of carbonation from the concrete surface because it changes from pink to colourless when the pH is less than 9.5 [6]. However, this method is not accurate enough to evaluate the durability of reinforced concrete, where the pH near the embedded steel is more than 11.5 [4,5]. A more accurate pH measurement method is needed to evaluate the durability performance of concrete materials, particularly for concrete containing supplementary cementitious materials (SCMs). Gruyaert et al. [7] reported that concrete containing SCMs is more vulnerable to carbonation than ordinary Portland cement concrete. Sanjuán et al. [8] also reported that mortars and concretes made of ground granulated blast-furnace slag (GGBFS) cement exhibited higher carbonation rates, particularly when they were poorly cured and had high GGBFS content.

Researchers have investigated the pH of CBMs for different purposes. Most studies are related to the passive protective film of embedded steel bars and aim to compare the pH among different CBM mixtures [9–11], to monitor the pH profiles over time [12] and at different depths [6,13–15], to determine the rates of pH changes [16] and the pH profiles of CBMs due to early carbonation curing [15,17]. Meanwhile, Kakade [18], and Räsänen and Penttala [19] measured the pH of a concrete surface for floor covering purposes. Alonso et al. [20] analysed methods for pH measurement for low-pH CBMs (pH about 11) used in a geological repository.

Few studies have used pH testing and phenolphthalein testing and determined their relation. Zhang and Shao [15] conducted phenolphthalein and pH measurement tests at different depths of the concrete surface at days 1 and 28 of different concrete mixtures, when the concrete samples were under early carbonation curing. At day 1, the pH test results at a carbonation depth of 10–25 mm showed higher variance than those at other depths of 0–10 and 25–50 mm. However, at day 28, no carbonation was found on the samples by using the phenolphthalein test, consistent with the pH test. The author suggested that this finding may be due to the existence of transition zones. McPolin [6] measured the apparent pH (not true pH) value of different carbonated concrete mixtures at different depths of the concrete surface. The measurements were conducted weekly, for a period of 6 weeks. At the 6th week, a phenolphthalein test was carried out to identify the average carbonation depth. The results showed that, at the average carbonated depth measured by phenolphthalein, the apparent pH value was within 11.3–11.7.

Behnood et al. [21] stated the lack of standardisation for pH measurement for concrete and urged the development of a profoundly reproducible and repeatable standard method for fresh and hardened concrete. Researchers typically applied their own consistent measurement methods across different samples for pH comparison (Figure 1). Pore solution extraction (PSE) method described in Barneyback and Diamond [22] is commonly referred by researches to extract pore solution from the CBMs for further analysis. Sagüés et al. [23] introduced in-situ leaching (ISL) method while ESL is the most widely applied method for pH measurement for CBMs. Each method has its advantages and disadvantages. However, there are insufficient studies about variables affecting each measurement method.

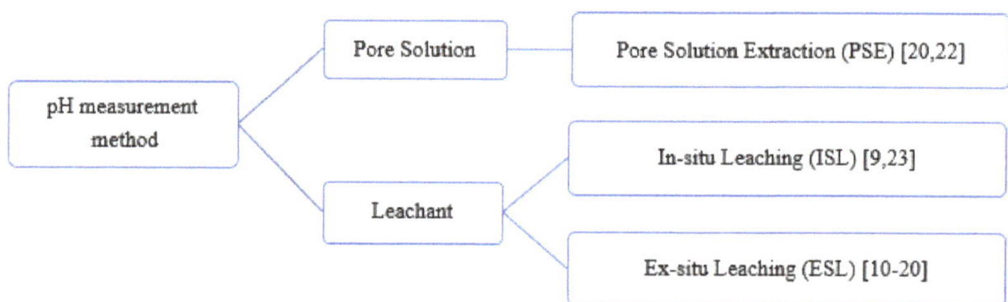

Figure 1. pH measurement methods categorised by different types of sample preparation.

Among different pH measurement methods, the ex-situ leaching (ESL) method is economic and simple [24]. ESL has good repeatability and reproducibility [20,25]. Figure 2 illustrates the steps and considerations for each stage of the ESL method; these parameters are inconsistent among researchers, varying from specimen preparation to measurement. Given the lack of a standardised method for ESL, experimental errors may occur due to the wrong selection of variables. As shown in this figure, the first step of measuring pH is to prepare fine particle sizes from the specimens. Table 1 lists the particle sizes used by different researchers [13,19,25,26] in their pH measurement.

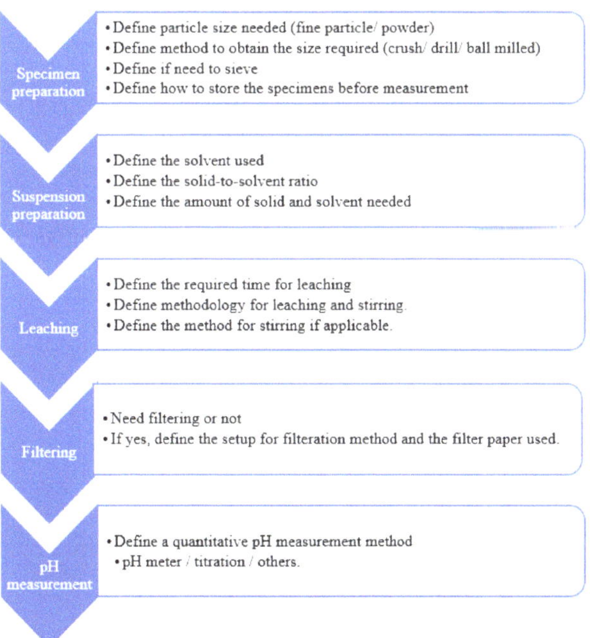

Figure 2. Summary for stages of the ESL method.

Table 1. Particle sizes used by different researchers in the ESL method.

Specimen Particle Size	Reference
Powder from drilling	McPolin et al. [6], Heng and Murata [16], Björk and Eriksson [27]
<63 μm	Rashad [10]
150 μm	Manso and Aguado [24]
<297 μm	Li et al. [28]
Grind until <80 μm, do not use a sieve	Alonso et al. [20]
Crush and grind using a vibratory disc mill at 1500 rpm for 30 s, then sieve to <80 μm	Plusquellec et al. [29]
<500 μm, 500 μm–1 mm, 1–2 mm	Pavlík [26]
<75 μm, 75–125 μm, 125–250 μm, 250–500 μm, 500 μm–1 mm, 1–2 mm, 2–4 mm, 4–8 mm, 8–16 mm, >16 mm	Räsänen and Penttala [19]
<850 μm, 850 μm–2 mm, >2 mm	Grubb et al. [13]
<74 μm, 74–149 μm, 149–297 μm, 297–595 μm	Wang et al. [25]

Pavlík [26] measured the concentration of OH^- ions (in mmol/L) for cement paste to calculate pH. The equation of pH = 14 + log[OH] was used in his study. The results showed that smaller particle sizes had a higher OH^- concentration, indicating a higher pH value (more alkaline). Räsänen and Penttala [19] tested concrete from blended cement with a solid-to-solvent ratio of 1:1, while Grubb et al. [13] tested concrete samples without fly ash, with a solid-to-solvent ratio of 1:2.

Table 2 shows the pH values of concrete in two different studies when different particle sizes were used. Specimens with bigger particle sizes had lower pH. Therefore, fine particles should be used for a more accurate measurement of pH [19]. However, preparing specimens into very fine particles requires more energy and time, compared with using bigger particle sizes.

Table 2. Comparison of the pH value for biggest and smallest particle sizes in previous studies.

Grubb et al. [13]		Räsänen and Penttala [19]	
Particle Size	pH Measured	Particle Size	pH Measured
>2 mm	11.55	>16 mm	12.22
<850 μm	12.48	<75 μm	12.81

Pavlík [26] reported that the effect of the particle size of Portland cement paste with a high solid-to-solvent ratio of 1:1 is relatively smaller than those with the ratios of 1:2, 1:5 and 1:50.

This study selected the ESL method to standardise a quantitative pH measurement method for CBMs to cover a wider pH range than phenolphthalein measurement. This work aims to determine the effects of using different particle sizes on the pH value of cement mortar and understand the possibility of using large particles (within the range of standard sand particle sizes) instead of using very fine particles for pH measurement. Similar studies were conducted on concrete and paste materials [13,19,25,26], and the results revealed that large particles had lower pH than very fine particles. Therefore, very fine particle sizes should be used to measure pH in the ESL method. No study has investigated the possibility of using large particles in pH measurement if very fine particles are not available. The use of large particles can benefit researchers and engineers with limited resources for making specimens in powdered form (very fine particle size). In addition, considering that the preparation of samples with large particle sizes is simpler, the time and cost of testing will be reduced.

2. Research Methodology

2.1. Materials

Low-porosity cement mortar with a cement to sand (c/s) ratio of 1:3 was prepared [30]. For a good workable mortar, a w/c ratio of 0.55 was applied. Ordinary Portland Cement (OPC) Type I (Tasek Corporation Berhad, Ipoh, Malaysia according to MS522 specification), local mining sand with a maximum grain size of 4.75 mm in saturated-surface-dry (SSD) condition and normal tap water from the pipeline were used. The fresh mortar, with a flow table test result of 220 mm, was cast in 50 mm cubes, and vibrated in two layers on a vibration table. The specimens were demoulded one day after casting and submerged in normal tap water until the time of testing (the age of specimens was 398 days at the time of testing). The average value of the compressive strength for the mortar sample was 55.7 MPa.

2.2. pH Meter Selection and Other Instruments

The main components of pH meter are an electrochemical system (a pH sensor and a reference electrode), a high input impedance meter (which processes the voltage difference in DC millivolt and displays them in mV or pH) and a temperature sensor. The reference electrode provides stable and constant reference voltage but is completely insensitive to the measured fluid. The junction of the reference electrode may be contaminated under adverse condition, such as high temperature, high pressure, high acidity and high basicity. An under-performing reference electrode can cause the pH reading to drift. Therefore, the reference electrode should be used carefully, according to the supplier's recommendation, and/or a double-junction electrode should be selected, instead of a single-junction electrode, to minimise contamination

The quality of electrodes degrades overtime depending on usage and maintenance, as evident in response time, slope and offset of the pH electrode. Response time refers to the time taken to obtain a stable reading. In theory, a perfect pH electrode reads 0 mV at pH 7 buffer at 25 °C, with a slope of 59.16 mV/pH. For a clean, correctly calibrated, and in good condition electrode, at 25 °C, the slope shall be within 92% to 110%, where 100% is 59.16 mV/pH, while the offset shall be $\pm$30 mV [31]. Before the test, calibration was performed with at least one to five standard buffer solutions, depending on the capability of the pH meter.

The suspension of CBMs consists of hard and sharp particles. The pH measurement was carried out in filtered aqueous solutions to protect the pH sensitive glass sensor from physical damage. The Hanna Instrument [32] recommends the use of plastic beakers to minimise any EMC interferences during the measurement.

This experiment used the Hanna edge® pH HI2002 and Hanna digital pH electrode HI11310, with a resolution of 0.01 pH and accuracy of $\pm$0.01 pH. The instrument has an automatic temperature compensation feature between -5 °C and 100 °C because of its built-in temperature sensor. This double-junction electrode has a 9.5 mm spherical tip.

2.3. Experimental Procedure

Test cubes were left to surface dry for about 15 min and crushed into pieces that could pass through sieve mesh no. 4 (4.75 mm). Throughout the test, CO_2 contamination on the specimens was minimised by keeping a distance from the operator's exhalation or other sources [23]. Wang et al. [25] reported insignificant pH difference (standard deviation of $\pm$0.03) among paste powder samples exposed to air for 0, 1, 5, 10, 20, 30, 60 and 120 min before mixing with deionised water. The specimens and liquid used were maintained at room temperature throughout the test. The specimens for pH measurement were sieved to obtain at least 50 g for each category as shown below:

(1) Mesh no. 8—to obtain particle size between 2.36 and 4.75 mm diameter;
(2) Mesh no. 16—to obtain particle size between 1.18 and 2.36 mm diameter;
(3) Mesh no. 30—to obtain particle size between 600 µm and 1.18 mm diameter;
(4) Mesh no. 50—to obtain particle size between 300 and 600 µm diameter;
(5) Mesh no. 100—to obtain particle size between 150 and 300 µm diameter;
(6) Base tray—to obtain particle size between less than 150 µm diameter.

Grubb et al. [13] and Plusquellec et al. [29] milled the specimens to measure the pH of CBMs. For comparison, about 100 g of crushed specimens were ground using planetary ball mills. The majority of these specimens passed through the 150 µm sieve after the milling process. Figure 3 displays the specimens with different sizes for the test. Fine powder appeared lighter in colour than large particles.

Each category of the specimens was stored in individual, sealed plastic bags. Before the test, the pH meter was calibrated with standard buffer solutions of 4.01, 7.01 and 10.01, with an offset of $\pm$30 mV and slope ranging between 92% and 110%.

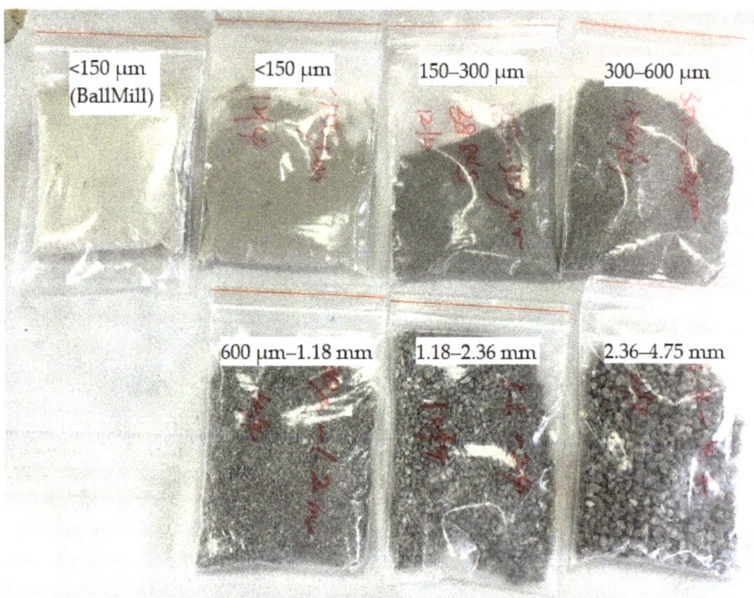

Figure 3. Samples stored in individual, sealed plastic bags.

In a 50 mL self-standing centrifuge tube, 10 ± 0.1 g of the specimens was added to 20 ± 0.1 g of fresh distilled water, followed by magnetic stir bar, before tightening the screw cap on the tube. The tube was placed on the magnetic stirrer to stir without heating for 5 min of leaching. The suspension was filtered through a quantitative ashless filter paper Grade 40 by using a Buchner funnel filter to obtain at least 10 mL of the solution. The filtered solution was immediately poured into a clean 50 mL self-standing centrifuge tube for pH measurement. The solution level was higher than the ceramic junction and stirred during the measurement. The meter was set to log the reading in stability-accurate mode to avoid human error. For each sample, three readings were recorded, and the mean value was taken. The test was conducted randomly until three samples were tested for each particle size. The estimated time for the experiment (after the specimens were separated into respective sizes) was as follows: preparation of the suspension: within 5 min; stirring and leaching: 5 min; filtering: within 1 min and pH measurement: within 5 min.

Figure 4 summarises the variables set for the benchmark methodology to test the effect of different particle sizes. Alonso et al. [20] and Wang et al. [25] applied 5 min of mixing with magnetic stirrer for leaching 10 g of powder, with the solid-to-solvent ratio of 1:1. The solid-to-solvent ratio of 1:2 for the benchmark methodology was set instead of 1:1, given that Grubb et al. [13] commented that this ratio was practical for the measurement. In brief, 20 g of water was chosen so that the filtered solution in the 50 mL tube was higher than the ceramic junction of the electrode during pH measurement. The suspension was filtered with Grade 40 filter paper to protect the electrode from physical damages [13,29]. A Bushner funnel was also used to prevent carbonisation during filtering.

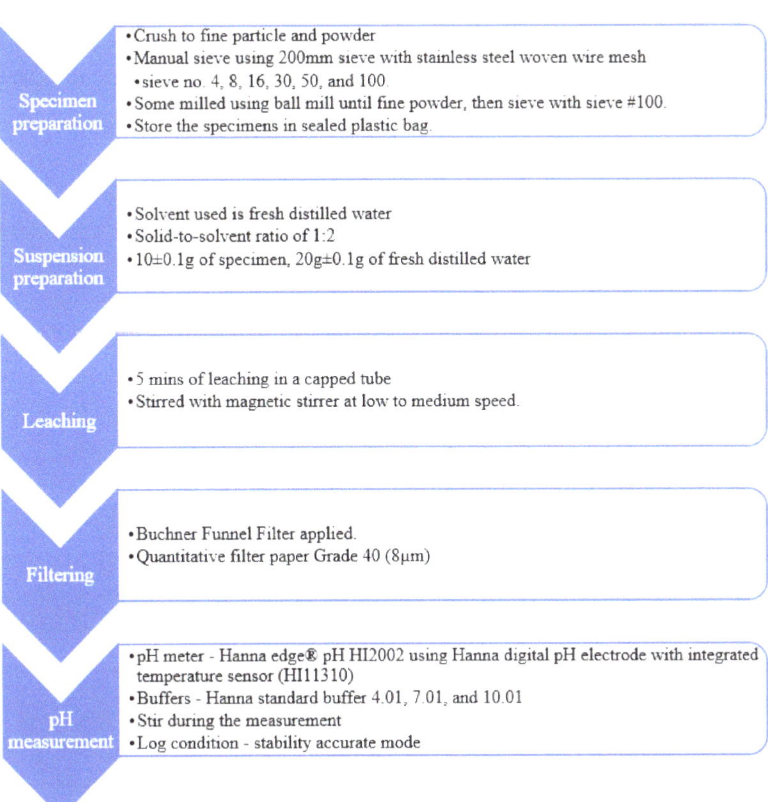

Figure 4. Benchmark methodology for effect of particle size.

3. Results and Discussion

3.1. Effect of Particle Size on pH

When a solid particle breaks into smaller particles, the total surface area increases. A large surface area allows more chemical reactions than a small surface area. Therefore, CBMs with smaller particle sizes have a higher total surface area in contact with solvent for the leaching of the OH^- ion [33]. Figure 5 shows that larger particle sizes have a lower pH value, as evident in particle sizes of 2.36–4.75 mm to 0.60–1.18 mm. For particle sizes of 2.36–4.75 mm, the pH is about 0.26 units lower than very fine particles (<150 μm using ball mill). Fine particles with sizes below 600 μm have a pH differential of less than 0.05 units. For pH measurement, the sample particle sizes could be divided into two zones, namely, large particle size (particle sizes larger than 600 μm) and fine particle size (particle sizes smaller than 600 μm). Figure 5 also shows that the proposed pH measurement method has good repeatability, as the standard deviations are 0.05 or less for each category of particle size, as can be seen in Table 3. The standard deviation for each particle size denotes that the electrode probe has high accuracy during the measurement.

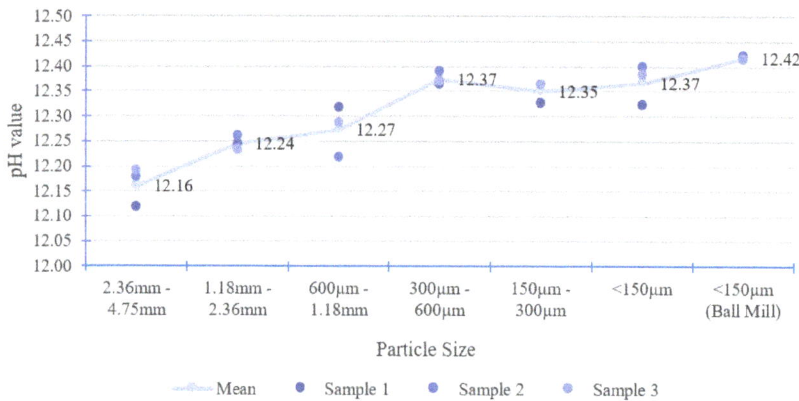

Figure 5. Effect of particle size.

Table 3. Comparison for pH among mortar, paste and concrete samples with similar particle sizes.

Particle Size	pH Value		
	In This Study (Mortar)	Wang et al. [25] (Paste)	Räsänen and Penttala [19] (Concrete)
2.36–4.75 mm	12.16 ± 0.04	N.A.	12.53
1.18–2.36 mm	12.24 ± 0.02	N.A.	12.64
600 μm–1.18 mm	12.27 ± 0.05	N.A.	12.72
300–600 μm	12.37 ± 0.02	12.73	12.74
150–300 μm	12.35 ± 0.02	12.73	12.77
<150 μm	12.37 ± 0.04	12.77	12.79

The results for cement mortar are consistent with the findings from other research (for similar particle size) but differ for CBMs, such as cement paste and concrete (Table 3). The pH value in the reference paper is higher than that in the present study, possibly due to the different calibration buffers used; this discrepancy requires further study for clarification. Wang et al. [25] determined the pH of cement paste by using calibration buffer solutions of 4, 7 and 10. Räsänen and Penttala [19] studied the pH of concrete by using calibration buffers of 10, 11, 12 and 13. The table also shows the similar trends of pH values for large and fine particle sizes.

3.2. Possibility of Using Samples with Large Particle Sizes in Measuring pH

Grubb et al. [13] reported that particle size influences the pH of CBMs, such that large particles showed lower pH than fine particles. Räsänen and Penttala [19] recommended that the particle size of samples should be fine or very fine to obtain an accurate result for the pH of cement-based materials; however, preparing fine particles is more difficult than large particles. Given the idea of using large particles for pH measurement, the following variables were examined.

3.2.1. Effect of Solid-to-Solvent Ratio on pH

The effect of solid-to-solvent ratio for large particles was determined. Large particle sizes refer to particles larger than 600 μm, while fine particle sizes refer to particles smaller than 600 μm. The solid-to-solvent ratio was changed from 1:2 to 1:1, where 20 g of solid was mixed with 20 g of fresh distilled water. On the same day, and with the same batch of specimens, samples with a size of 300–600 μm were tested according to the benchmark methodology for reference (Figure 2). The measured pH of 12.35 ± 0.02 matched the results reported in Section 3.1.

Table 4 shows the pH test results of three different ranges of large particle sizes (large particle sizes in ranges I, II and III) with the solid-to-solvent ratio of 1:1 in comparison with the pH value of fine particle sizes with the solid-to-solvent ratio of 1:2. With increasing solid-to-solvent ratio of the large particle sizes, the pH value of the range III particle size is similar to the reference, whereas that of the range II particle size differs from the reference. A meaningful difference was found between the range I particle size and the reference. By increasing the quantity of solid in the same amount of solvent, the total surface area increases, and a higher pH value (equal or close) than the reference could be achieved.

Table 4. Effect of the solid-to-solvent ratio of different large particle sizes on pH.

Particle Size	Large Particle Size			Fine Particle Size
	2.36–4.75 mm (Particle Size Range I)	1.18–2.36 mm (Particle Size Range II)	600 μm–1.18 mm (Particle Size Range III)	300–600 μm (Reference)
solid-to-solvent ratio	1:1	1:1	1:1	1:2
pH value (mean ± SD)	12.23 ± 0.02	12.31 ± 0.04	12.35 ± 0.02	12.35 ± 0.02

Figure 6 shows the pH values of the three ranges of large particle size with different solid-to-solvent ratios. The pH increased by about 0.07 units when the quantity of the solid samples was doubled for all the three ranges of the large particle size. Grubb et al. [13] reported that when the solid-to-solvent ratio of ground cement paste was changed from 1:2 to 1:1 (with 20 g of water), the pH increased by about 0.05 units. Räsänen and Penttala [19] tested the solid-to-solvent ratio for 7-day-old normal-strength concrete (NSC-7) and 180-day-old high-strength concrete (HSC-180), which were ground into powder. When the solid-to-solvent ratio was increased from 1:2 to 1:1, the pH value increased by approximately 0.04 units for NSC-7 and 0.12 units for HSC-180. The change in the solid-to-solvent ratio from 1:2 to 1:1 did not contribute much to the pH increment compared with the large particle sizes tested.

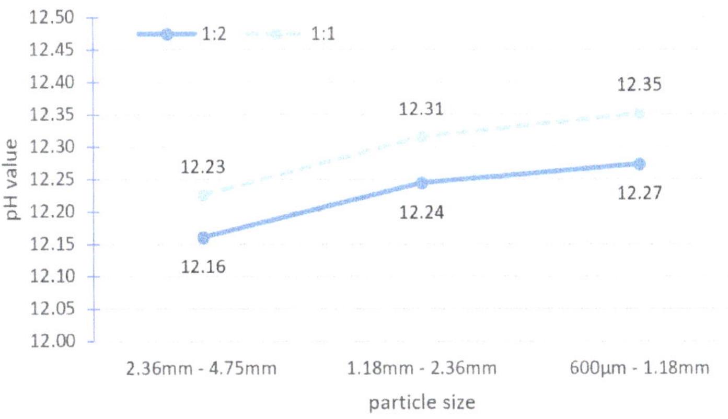

Figure 6. Effect of solid-to-solvent ratio of 1:1 and 1:2 for large particle size.

3.2.2. Effect of Leaching Time on pH

The leaching time was increased to investigate the possibility of using large particle sizes for accurate pH measurement. The specimens were placed in a magnetic stirrer for 10 min of leaching instead of 5 min. The reference sample was treated similarly to the method in Section 3.2.1, and the solid-to-solvent ratio for all specimens was 1:2. The test results are shown in Table 5. With increasing leaching time, the pH also increases, indicating that the leaching time for the benchmark methodology (Figure 4) may not be sufficient when the particle size of the specimens was large. The pH values of the ranges I and II large particle sizes are almost the same as the reference sample, but higher than that of the range III large particle size.

Table 5. Effect of leaching time for different large particle sizes on pH.

Particle Size	Large Particle Size			Fine Particle Size
	2.36–4.75 mm (Particle Size Range I)	1.18–2.36 mm (Particle Size Range II)	600 µm–1.18 mm (Particle Size Range III)	300–600 µm (Reference)
Leaching time (min)	10	10	10	5
pH value (mean ± SD)	12.32 ± 0.03	12.38 ± 0.02	12.43 ± 0.03	12.35 ± 0.02

Figure 7 shows an average increment in the pH value of about 0.15 for large particle sizes when the leaching time is increased from 5 min to 10 min. Räsänen and Penttala [19] utilised a CM200 cell mixer with the speed of 35 rpm to stir the suspension. The stirring speed is slower than that of magnetic stirring used in the present study. For NSC-180, when the stirring time is increased from 5 min to 15 min, the pH increases by 0.01 units. Grubb et al. [13] recorded an insignificant pH change of approximately 0.05 units for a leaching time of 30–60 min. In both studies, the pH tests were conducted on the fine particle size (powder form).

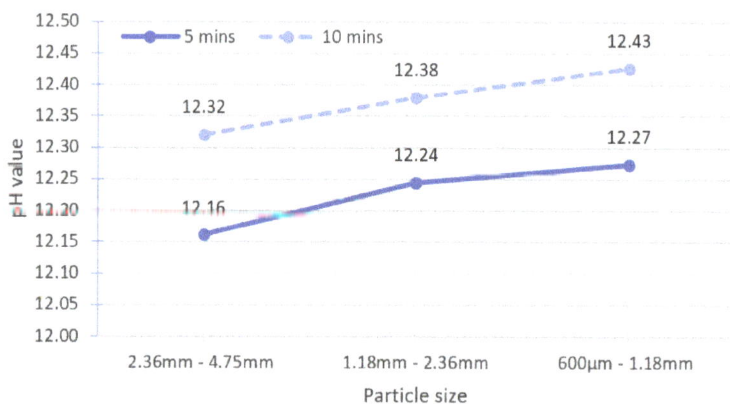

Figure 7. Effect of changing leaching time to 10 min for large particle size.

4. Conclusions

As a part of durability and serviceability assessment of a concrete structure, the pH of structural concrete is one of the fundamental indicators to ensure that the concrete structure is in healthy condition or assess whether it needs rehabilitation and maintenance. Considering the lack of a standard for pH measurement, especially for high pH ranges, this work investigated the possibility of using large particle sizes, instead of fine particle sizes, for CBMs (cement mortar) to simplify the method of pH measurement.

Cement mortar specimens with smaller particle sizes resulted in a higher pH value, consistent with the studies on cement paste and concrete. Using a particle size of less than 600 µm (fine particle size) showed consistent pH measurement with a mean of 12.37–12.42 for cement mortar. In addition, large particle sizes (particle sizes between 2.36 mm and 600 µm) can be used for accurate pH measurement if the solid-to-solvent ratio is increased from 1:2 (used for fine particle sizes) to 1:1 or the leaching time is doubled (from 5 min for fine particle sizes to 10 min). However, increasing the leaching time is recommended, rather than increasing the solid-to-solvent ratio to obtain similar pH readings as fine particle size.

These findings are important for the development of a standardised pH measurement using the ESL method for CBMs. The effect of other variables should be investigated on

different types of CBMs to verify the applicability of all variables before a standardised pH measurement can be proposed.

Author Contributions: Conceptualization, P.Y.L. and P.S.; methodology, P.Y.L. and P.S.; software, P.Y.L.; validation, P.Y.L., P.S., H.Y.B.K., Z.I. and S.Y.; formal analysis, P.Y.L. and P.S.; investigation, P.Y.L.; resources, P.Y.L.; data curation, P.Y.L. and P.S.; writing—original draft preparation, P.Y.L.; writing—review and editing, P.Y.L., P.S., H.Y.B.K., Z.I. and S.Y.; visualization, P.Y.L.; supervision, P.S. and Z.I.; funding acquisition, P.S. and H.Y.B.K. All authors have read and agreed to the published version of the manuscript.

Funding: This research is supported by Universiti Tenaga Nasional Internal Research Grant OPEX (J5100D4103-BOLDREFRESH2025-CENTRE OF EXCELLENCE) and Universiti Malaya research grant (Grant No. GPF007F-2019).

Institutional Review Board Statement: Not applicable.

Informed Consent Statement: Not applicable.

Data Availability Statement: The data presented in this study are available within the article.

Acknowledgments: The authors would like to express their gratitude for the provided support by Universiti Malaya and Universiti Tenaga Nasional.

Conflicts of Interest: The authors declare no conflict of interest.

References

1. Neville, A.M. Portland cement. In *Properties of Concrete*, 5th ed.; Pearson Education Limited: London, UK, 2011.
2. Li, Z.J. Materials for making concrete. In *Advanced Concrete Technology*; John Wiley & Sons. Inc.: Hoboken, NJ, USA, 2011; p. 38.
3. ACI Committee 222. *Protection of Metals in Concrete Against Corrosion*; American Concrete Institute: Farmington Hills, MI, USA, 2001.
4. Taylor, H.F.W. Concrete chemistry. In *Cement Chemistry*; Academic Press Limited: London, UK, 1990; pp. 383–387.
5. Kurdowski, W. Concrete Properties. In *Cement and Concrete Chemistry*; Springer: Dordrecht, The Netherlands, 2014; p. 478. [CrossRef]
6. McPolin, D.O.; Bashneer, P.A.M.; Long, A.E.; Grattan, K.T.V.; Sun, T. New test method to obtain pH profiles due to carbonation of concretes containing supplementary cementitious materials. *J. Mater. Civ. Eng.* **2007**, *19*, 936–956. [CrossRef]
7. Gruyaert, E.; Van Den Heede, P.; De Belie, N. Carbonation of slag concrete: Effect of the cement replacement level and curing on the carbonation coefficient—Effect of carbonation on the pore structure. *Cem. Concr. Comp.* **2013**, *35*, 39–48. [CrossRef]
8. Sanjuán, M.A.; Estévez, E.; Argiz, C.; Del Barrio, D. Effect of curing time on granulated blast-furnace slag cement mortars carbonation. *Cem. Concr. Comp.* **2018**, *90*, 257–265. [CrossRef]
9. Li, L.F.; Sagüés, A.A.; Poor, N. In situ leaching investigation of pH and nitrite concentration in concrete pore solution. *Cem. Concr. Res.* **1999**, *29*, 315–321. [CrossRef]
10. Rashad, A.M. An investigation on very high volume slag pastes subjected to elevated temperatures. *Constr. Build. Mater.* **2015**, *74*, 249–258. [CrossRef]
11. Shafigh, P.; Yousuf, S.; Lee, C.J.; Ibrahim, Z. The effect of cement mortar composition on the pH value. *IOP Conf. Ser. Mater. Sci. Eng.* **2020**, *770*, 012026. [CrossRef]
12. Gowripalan, N.; Mohamed, H.M. Chloride-ion induced corrosion of galvanized and ordinary steel reinforcement in high-performance concrete. *Cem. Concr. Res.* **1998**, *28*, 1119–1131. [CrossRef]
13. Grubb, J.A.; Limaye, H.S.; Ashok, M.K. Testing pH of Concrete. *Concr. Int.* **2007**, *29*, 78–83.
14. Rostami, V.; Shao, Y.X.; Boyd, A.J. Durability of concrete pipes subjected to combined steam and carbonation curing. *Constr. Build. Mater.* **2011**, *25*, 3345–3355. [CrossRef]
15. Zhang, D.; Shao, Y.X. Effect of early carbonation curing on chloride penetration and weathering carbonation in concrete. *Constr. Build. Mater.* **2016**, *123*, 516–526. [CrossRef]
16. Heng, M.; Murata, K. Aging of concrete buildings and determining the pH value on the surface of concrete by using a handy semi-conductive pH meter. *Anal. Sci.* **2004**, *20*, 1087–1090. [CrossRef] [PubMed]
17. Rostami, V.; Shao, Y.X.; Boyd, A.J.; He, Z. Microstructure of cement paste subject to early carbonation curing. *Cem. Concr. Res.* **2012**, *42*, 186–193. [CrossRef]
18. Kakade, A.M. Measuring concrete surface pH—A proposed test method. *Concr. Repair Bull.* **2014**, 16–20. Available online: https://cdn.ymaws.com/www.icri.org/resource/resmgr/crb/2014marapr/CRBMarApr14_Kakade.pdf (accessed on 16 June 2021).
19. Räsänen, V.; Penttala, V. The pH measurement of concrete and smoothing mortar using a concrete powder suspension. *Cem. Concr. Res.* **2004**, *34*, 813–820. [CrossRef]

20. Alonso, M.C.; Calvo, J.L.G.; Walker, C.; Naito, M.; Pettersson, S.; Puigdomenech, I.; Cuñado, M.A.; Vuorio, M.; Weber, H.; Ueda, H.; et al. *Development of an Accurate pH Measurement Methodology for the Pore Fluids of Low pH Cementitious Materials*; SKB R-12-02; Swedish Nuclear Fuel and Waste Management Co.: Stockholm, Sweden, 2012.
21. Behnood, A.; Van Tittelboom, K.; De Belie, N. Methods for measuring pH in concrete: A review. *Constr. Build. Mater.* **2016**, *105*, 176–188. [CrossRef]
22. Barneyback Jr, R.; Diamond, S. Expression and analysis of pore fluids from hardened cement pastes and mortars. *Cem. Concr. Res.* **1981**, *11*, 279–285. [CrossRef]
23. Sagüés, A.A.; Moreno, E.I.; Andrade, C. Evolution of pH during in-situ leaching in small concrete cavities. *Cem. Concr. Res.* **1997**, *27*, 1747–1759. [CrossRef]
24. Manso, S.; Aguado, A. A review of sample preparation and its influence on pH determination in concrete samples. *Mater. Construcción* **2017**, *67*, 1–10. [CrossRef]
25. Wang, W.C.; Huang, W.H.; Lee, M.Y.; Duong, H.T.H.; Chang, Y.H. Standardized Procedure of Measuring the pH Value of Cement Matrix Material by Ex-Situ Leaching Method (ESL). *Crystals* **2021**, *11*, 436. [CrossRef]
26. Pavlík, V. Water extraction of chloride, hydroxide and other ions from hardened cement pastes. *Cem. Concr. Res.* **2000**, *30*, 895–906. [CrossRef]
27. Björk, F.; Eriksson, C.A. Measurement of alkalinity in concrete by a simple procedure, to investigate transport of alkaline material from the concrete slab to a self-levelling screed. *Constr. Build. Mater.* **2002**, *16*, 535–542. [CrossRef]
28. Li, L.F.; Nam, J.; Hartt, W.H. Ex situ leaching measurement of concrete alkalinity. *Cem. Concr. Res.* **2005**, *35*, 277–283. [CrossRef]
29. Plusquellec, G.; Geiker, M.; Lindgård, J.; Duchesne, J.; Fournier, B.; De Weerdt, K. Determination of the pH and the free alkali metal content in the pore solution of concrete: Review and experimental comparison. *Cem. Concr. Res.* **2017**, *96*, 13–26. [CrossRef]
30. Shafigh, P.; Asadi, P.; Akhiani, A.R.; Mahyuddin, N.B.; Hashemi, M. Thermal properties of cement mortar with different mix proportions. *Mat. Constr.* **2020**, *70*, 224. [CrossRef]
31. Hanna Instruments. pH. In *General Catalog*; Hanna Instruments Inc.: Woonsocket, RI, USA, 2020; Volume 34, pp. 2.1–2.170.
32. Hanna Instruments. Operational guide. In *edge®pH Instruction Manual*; Hanna Instruments Inc.: Woonsocket, RI, USA; p. 37. Available online: https://www.manualslib.com/manual/1555430/Hanna-Instruments-Edge.html#manual (accessed on 15 June 2021).
33. Meyers, R.A. *Encyclopedia of Environmental Analysis and Remediation. 8 Volume Set*; Wiley-Interscience: New York, NY, USA, 1998; p. 2094.

MDPI

St. Alban-Anlage 66

4052 Basel

Switzerland

Tel. +41 61 683 77 34

Fax +41 61 302 89 18

www.mdpi.com

Applied Sciences Editorial Office

E-mail: applsci@mdpi.com

www.mdpi.com/journal/applsci

www.ingramcontent.com/pod-product-compliance
Lightning Source LLC
LaVergne TN
LVHW070650170726
843515LV00004B/371